THE PROCESS OF RELIABILITY ENGINEERING

THE PROCESS OF RELIABILITY ENGINEERING

Creating Reliability Plans
That Add Value

Carl S. Carlson
and
Fred Schenkelberg

Publishing
FMS Reliability Publishing
Los Gatos, California

To order copies or permission requests write to the publisher:

FMS Reliability Publishing
17810 Comanche Trail
Los Gatos, CA 95033
fmsreliability.com/publishing
fms@fmsreliability.com

Printed in the United States of America

Library of Congress Control Number: 2023901278

ePub ISBN: 978-1-938122-10-1
Paperback ISBN: 978-1-938122-12-5

Book Layout © 2021 BookDesignTemplates.com
Book Cover by Andy Meaden
Editing by David Couzens
Typeset in Garamond Premier Pro

Publisher's Cataloging-in-Publication
(Provided by Cassidy Cataloguing Services, Inc.).

Names: Carlson, Carl S. (Carl Seymour), author. | Schenkelberg, Fred, author.
Title: The process of reliability engineering : creating reliability plans that add value / Carl S. Carlson
            and Fred Schenkelberg.
Description: Los Gatos, California : FMS Reliability Publishing, [2023] | Includes bibliographical
            references.
Identifiers: ISBN: 978-1-938122-12-5 (paperback) | 978-1-938122-10-1 (ePub) | LCCN:
            2023901278
Subjects: LCSH: Reliability (Engineering) | Reliability (Engineering)--Management. | Project
            management. | Factory management. | Industrial management. | Strategic planning. |
            Decision making. | BISAC: BUSINESS & ECONOMICS / Management Science. |
            TECHNOLOGY & ENGINEERING / Quality Control. | TECHNOLOGY & ENGINEERING /
            Systems Engineering.
Classification: LCC: TS173 .C37 2023 | DDC: 620/.00452--dc23

DEDICATION

To my dear Holly, whose never-ending support has been the very foundation of my writing, and who has taught me to always consider the audience in everything I write. C.S.C.

To Diane, my adventures partner, best friend, and spouse, for her patience and encouragement, and the many reminders to sit down and write. F.S.

Destiny is no matter of chance. It is a matter of choice.
It is not a thing to be waited for, it is a thing to be achieved.
—William Jennings Bryan

Contents

PREFACE

When we began writing a book about reliability management, it dawned on us that there are plenty of books and resources on how to conduct failure mode and effects analysis (FMEA), accelerated life tests, and failure analysis and explaining the rest of the ways we work. The methods of reliability engineering are well documented. What is missing, in our opinion, is when and how to use those methods well. How do you put it all together in such a way that one is likely to obtain the desired results?

We both have long known that performing reliability activities, such as facilitating an FMEA or analyzing data from an accelerated life test, is just one element of managing reliability within an organization. The culture in an organization concerning reliability decision-making and how reliability is thought about and prioritized is more important than a well-fit Weibull curve. Part of managing reliability is managing the culture, which means focusing on how decisions are made. It means creating processes that enhance our ability to ensure that each decision includes the appropriate reliability information and priority.

Influencing the culture is done by creating and executing one reliability plan after another that achieves the specific project's reliability objectives and moves the organization's culture to incorporate reliability activities and thinking into everyone's routine work. To influence the culture, one first has to be able to influence at least one other person, then small groups, then the entire team. Our ability to influence finds footing with our technical and analytical skills, plus our ability to communicate.

In our experience, it is possible to change an organization's culture concerning reliability. Also, from our experience, we know it is not a simple or quick process. Changing how a group of people asks questions, seeks information, and makes decisions is often a reflection of respected leaders within the organization. To improve the ability of a team to value reliability engineering methods, they need to know what the various methods are, how to use those methods, and why they are useful. This takes experience—which takes time.

Creating trust in the recommendations and plans within a reliability plan ensures that each task has a purpose and adds value. Each task has to make a difference. It is the accumulation of meaningful results that builds trust in the methods. Being an effective communicator certainly helps, which we also cover throughout this book.

Creating reliability plans that achieve reliability objectives and improve the reliability maturity of the organization takes both technical and communication skills. One without the other is insufficient to create lasting change. When done well, applying the right methods with the full support and encouragement of the entire team, the changes to the culture will continue, and the organization will be able to create an effective reliability plan and continue to improve the culture around reliability long after you have left.

We've successfully crafted reliability plans that change how organizations approach creating reliable products, and so can you. We are happy to share our experience in this work so you can be successful. We sincerely hope this book provides each reader with insights and tools to reach your reliability goals. We wish you the best in your quest to support safe and trouble-free products and processes.

December 4, 2022

ACKNOWLEDGMENTS

The work would not be possible if not for the many opportunities and experiences our careers have provided. We are both grateful to have learned from so many.

We are in awe of the hundreds of individuals that took time to read and comment on early drafts. This book took shape and improved with each identified spelling error, topic suggestion, and insightful observation.

Special thanks are owed to Greg Hutchins, Philip Sage, Chris Jackson, Caroline Lubert, John Paschkewitz, Arthur Hart, Dev Raheja, Andre Kleyner, William Meeker, and Les Warrington for their support, insights, encouragement, and guidance.

INTRODUCTION

Reliability occurs at the point of decision.

Why this book is useful

We have witnessed organizations that consistently create highly reliable products. —along with plenty of organizations with inconsistent results. The difference is not the presence of a reliability team; rather, it is the careful selection of reliability methods that add value. Selecting and executing reliability tasks that provide the necessary information or insights that enable better decision-making is the key. We have worked with dozens of organizations to help them learn how to achieve their desired reliability results consistently.

This book guides us through creating an organization that consistently creates highly reliable products. Although those with reliability engineering experience bring skills and knowledge that may help, they are not necessary. The techniques, tools, and methods used and referred to in this book are not the sole domain of reliability engineering, thus any product development team can and should know and use these tools directly. Of course, a reliability engineer on staff may help the rest of the team learn to master and incorporate reliability methods into regular use.

This book focuses on the process of crafting reliability plans that assist in the creation of products or systems that meet business and customer objectives. This book is not about the details of how to conduct a design failure mode and effects analysis (FMEA) or perform accelerated life testing (ALT). Instead, it is about when and why to use these methods.

The available standards and guidelines do not include the details on how to select the right set of tools to maximize the plan's likelihood to create a reliable product in a resource-efficient manner. This book builds on the existing literature by including details on understanding the current capabilities and limitations. Moreover, there is a focus on identifying key decisions and detailing a process to actually select the most suitable reliability methods for the unique circumstances.

The trick to creating a product that meets the reliability performance objectives and improves the ability of the team to do so consistently is not well covered in other works. Understanding how the blend of capabilities and behaviors creates a culture within an organization related to reliability provides a starting point. Changing a culture is not easy work, yet it is possible, as we have done so in working with numerous organizations. By starting with a good plan, coupled with using excellent communication skills, it is possible for you to do so also.

Reliability plan standards and guidelines

There are several published standards and guidelines that cover the scope and general procedure for reliability plans. Standards and guidelines provide a common language to use when working to create a product or system. They also provide a range of information about specific methods that may assist in creating a reliable product.

Industry standards and guidelines are not templates for direct use as a reliability plan. Most of these documents include directions such as in MIL-STD-785B (1980, p. iii):

Effective reliability programs must be tailored to fit program needs and constraints, including life cycle costs (LCC). This document is intentionally structured to discourage indiscriminate blanket applications. Tailoring is forced by requiring that specific tasks be selected...

Many available documents provide guidance on the use of a wide range of reliability methods. They often include the necessary conditions for effective use and expected outputs for each method. Yet what is missing is they do not provide adequate guidance to prioritize and select the vital few methods that will achieve reliability objectives, given the resources and capabilities of the organization. The process detailed in this book bridges this gap. As readers, you will learn how to select, organize, and implement an effective reliability plan. You will learn a proven process to achieve high reliability.

Here are some of the more common and relevant guidelines:

- TAHB0009A, Reliability Program Handbook, May, 2019, SAE International.
- GEIA-STD-0009A, Reliability Program Standard for Systems Design, Development, and Manufacturing, May 2020, SAE International.
- JA1003_201205, Software Reliability Program Implementation Guide, May 2012, SAE International.
- MIL-STD-785B, Reliability Program for Systems and Equipment Development and Production, September 1980, U.S. Department of Defense.
- IEC 60300-1 Ed. 3.0, Dependability Management—Part 1: Guidance for Management and Application, January 2012, IEC TC 56 Committee.
- IEEE 1332, Standard Reliability Program for the Development and Production of Electronic Systems and Equipment, 2012.
- DoD Guide for Achieving Reliability, Availability, and Maintainability, August 2005, U.S. Department of Defense.

Example scenarios used throughout the book

To illustrate the difference between the various sets of priorities and constraints, we are creating three fictitious bicycle product line scenarios for use throughout the book.

Pro Series (low-volume, high-cost scenario)

The Pro Series bicycle is designed for road bike racing teams and serious amateurs. Pro Series bicycles are custom designed with the latest in materials, technology, and artisanship to exacting standards of professional cycling teams. They are directly sold to professional cycling teams and select high-end bicycle specialty shops. Weight, responsiveness, and overall performance are critical factors. Cost is generally not an issue. Long-term reliability considerations are minimized, as teams generally use new bicycles each season. Serious amateurs may use the equipment for two or three years. However, reliability and safety during the shorter intended life of the bicycle are very important. Very harsh use, high power, environmental conditions, daily pressure-washing, and tune-ups stress every bicycle component.

Very low volumes of custom bicycles are manufactured for professional teams—perhaps 50–200 per year. The best and most popular features are incorporated into the retail version at a very high per frame price point, with production runs of maybe 7,000–10,000 units annually. The company's management focuses on design and performance, with 15% of net revenue invested into research and development. Nearly every staff member rides 300 miles a week. The lead designer is well known in the professional cycling world.

The design team attempts to reduce weight and increase performance for the Pro Series products. They strive to construct a bicycle that performs without failure (structural failure) by regularly considering accumulated damage-type failure mechanisms to balance the chance of failure and performance. A failed frame is life-threatening to highly trained and skilled professional riders.

Enthusiast Series(moderate-volume, moderate-cost scenario)

The Enthusiast Series bicycle is designed for road, triathlon, and mountain bike recreational riders. Bicycle enthusiasts ride for fun, recreation, and fitness. Retail outlets include nearly all bicycle shops and related specialty stores worldwide. Some ride in organized events or just on weekends with friends. Some commute with their bicycles. There are multiple families of relatively high-volume bicycle lines. Road, trip, and mountain lines each have various price points; these are accompanied with different accessories (derailleur, sprockets, brakes, seat, and wheels). Frames are generally based on the technology of pro bicycles from two or more prior seasons. Proven technology with high-volume production is possible. Environmental conditions and the range of use conditions are very wide. Road and trail conditions vary extensively. Bicycles are owned and ridden for 1–10 years; a few are regularly used much longer. Production quantities of 75,000–100,000 per base frame are configured with accessories into three price points. There are nine products in a typical season. A frame design may be in production for two or three years.

The Enthusiast Series division was purchased about a year ago and did not have an active reliability program. It filled the need for the growing company to offer bicycles at mid-range prices. The year since the purchase has illuminated the need to improve the reliability of every line. Warranty costs and brand image threaten the Enthusiast and Pro Series' reputation.

Intro Series (high-volume, low-cost scenario)

The Intro Series is designed for those new to cycling or who need an economical bicycle solution. Its bicycles are retailed at chain stores, most bicycle shops, and online worldwide. Volumes are 250,000–400,000 per year per model. There are six basic models: two each in recreational, road, and mountain bike product lines. Technology and accessories are selected for durability and price. Most bicycles are used for many years; a warranty of two years is common. The intended useful life is 5–10 years. Riders use these bicycles for transportation to work or school and for occasional recreational use for enjoyment. The bicycles receive a minimum of attention, shelter, and maintenance. The Intro Series division was purchased about five years ago as part of a strategy to provide a full line of bicycles.

The division has steadily improved reliability performance under the guidance of the parent company and its goals are to provide reliable bicycles at the lowest cost. The division relies on product testing to identify and improve the design. Field failures do occur, and focused teams address and make improvements. High volume has forced improvements in quality control yet often relies on testing.

Web companion to *The Process of Reliability Engineering*

Visit www.accendoreliability.com/go/pre to find the book's appendices and additional resources related to this work. Also available are downloadable and printable versions of the reliability maturity matrix and the six-step process to achieving high reliability, as well as updated references and recommendations.

We also would like to hear from you. The website will include a comment form plus contact information so that you can pass along your thoughts or questions. The work of reliability engineering continues to evolve, so the intent is to stay in touch with you, our readers, and continue to update, assist, and support your work as best we can. We look forward to hearing from you.

Chapter 1

MAKING THE CASE FOR EFFECTIVE RELIABILITY PLANS

The road to success is not a path you find but a trail you blaze.
Robert Breault

In this chapter. We will explain how well-constructed and executed reliability plans will achieve your organization's reliability goals and objectives. We present what we call the Reliability Challenge and describe the value, scope, and primary steps to achieving high reliability. We end with a few success stories.

1.1 Decisions, decisions, decisions

We recognize that reliability engineering is often about our ability to influence decisions such that the team or organization can consistently achieve the desired product reliability performance. Crafting a reliability plan is a process to understand objectives, capability, obstacles, and constraints so that we select and execute the right reliability methods. One criterion for the "right method" is that it makes a positive difference or adds value.

A common management tool is to create and execute a plan. The structure of a well-thought-out and informative plan enables an organization to realize an objective. For reliability engineering management, the same rule applies. To achieve the desired reliability outcomes for your business and customers requires creating and executing a reliability plan.

The intent of this book is to share a proven method to create and execute a reliability plan that helps your organization consistently achieve reliability objectives. Achieving your reliability goals is not simply about doing reliability-engineering-type activities. Rather, it is about understanding and overcoming the barriers to having the right information at the right time so that your organization can make the right decisions to balance reliability performance with other business objectives.

Each step in the process of creating highly reliable products adds value. The process adds the clarity of objectives and obstacles to understanding what needs to be done and why the selected methods are the most suitable ones. A reliability plan is a vehicle to inform decision makers as they work. Your product's reliability occurs through those decisions.

1.2 The reliability challenge

Today's corporations are facing unprecedented worldwide competition as a result of three continuing challenges: the mandate to reduce costs, the need for faster development times, and the expectation of high reliability for products and processes. The necessity for reliability assurance will not abate; however, there is increasing emphasis on design for reliability (DFR) as a meaningful corporate strategy. This means rethinking our approach to achieving high reliability.

Successful companies incorporate a reliability philosophy that supports the following principles:

- Senior management embraces an authentic reliability vision.
- A proactive reliability culture exists and is encouraged.
- Reliability is designed into products and processes.
- Use is made of the best available science-based methods.

Knowing how to calculate reliability is important. Knowing how to achieve reliability is equally, if not more, important. Design for reliability practices must begin early in the design process and be integrated into the overall product development cycle.

Simply stated, the reliability challenge is as follows:

Develop, manufacture, and service products that achieve impeccable customer satisfaction, trouble-free operation, and proven safety, and get them into the marketplace faster than the competition at the lowest possible cost.

1.3 Reliability engineering versus reliability management

The field of reliability has two primary disciplines: reliability engineering and reliability management. Reliability engineering is concerned with the knowledge and application of the specific tasks and methods that achieve high reliability, while reliability management focuses on how to effectively use reliability resources and select and implement the correct reliability tools to meet program safety and reliability objectives. Developing effective reliability plans relies upon using both disciplines.

In many companies, engineering departments in development, manufacturing, and operations are supported by some level of expert reliability engineering resources. Reliability engineers do not design, produce, or distribute products or systems yet work to enable others to make better decisions regarding the impact on reliability performance. These people have varying expertise in the subject matter of reliability tools and methods. It is a common mistake for management to believe that the program reliability goals will be met by assembling a team of expert reliability resources. Engineering staff must be supported by proactive planning and managing of reliability tasks.

This is not to imply a reliability person or team is required. The roles may reside within one or more of the engineering staff. The culture around the consideration of reliability aspects of each decision is key.

For example, in the case of accelerated life testing (ALT), management will identify the need and value of performing ALT and plan for the required resources, reliability engineers will perform an analysis of the ALT results and make appropriate recommendations, and management will follow up with executing the resulting design or process changes. Both disciplines are necessary for a successful reliability program.

1.4 Importance of influencing decisions

A well-written report that is not read is of no value. It may contain information that would prevent a major field problem, yet the report changes nothing if no one reads it. The same applies to recommendations, findings, results, estimates, etc. If others do not understand and use the information provided, the effort to create the recommendations is a waste of time.

Reliability occurs at the point of decision. Which vendor should you select? Which material do you use? Is this design reliable enough? Is this process doable? In addition, as with these and the myriad of decisions involved in the creation of a product, the impact on reliability performance is a part of the decision. Crafting recommendations, running tests, examining data, and writing reports are all examples of creating and providing information to influence decisions such that decision makers consider the impact on reliability.

Each step and task involved with creating a reliable product is an attempt to add value. By this, we mean that the investment and effort made to conduct reliability-related tasks or activities will return a benefit greater than the investment. A clear reliability objective provides a common goal and reduces wasted effort by working at cross-purposes on conflicting objectives. The selection of the appropriate reliability method involves, in part, choosing which method is likely to provide the right amount of information to influence decisions and result in lower field failure rates, saved engineering time, reduced chance of a project delay, or creating a benefit by one of many other potential sources of value. Each task in the reliability plan has a clear path to how the results will create value and influence the team to create a reliable product.

Those working on reliability-specific topics are only successful if they can influence decisions with the appropriate information at the right time. A reliability plan is a roadmap to identify and influence decisions across the product life cycle. A plan that is not implemented and in which the outputs of the various activities are not incorporated into decision-making is not worth the effort.

Influence is based, in part, on technical skill. Yet, skill alone is insufficient in most cases. The professional also needs to have the trust of the decision maker. It is possible to improve your ability to influence others. This involves technical prowess as well as excellent communication and social skills.

1.5 The need for effective reliability plans

In the book *Handbook of Reliability Engineering and Management*, Bradin (1996, p. 2.9) writes the following:

> To be effective, reliability policy must issue from high-level management. Management's attitude toward reliability, expressed through policy, is the most important single ingredient in making reliability engineering and reliability assurance a successful practice in any organization.

Management influences the reliability culture in a company. It is not a slogan, but the way people interact, behave, and make decisions that forms the culture. In a nutshell, effective reliability planning saves time and resources and improves reliability and business results. This book will outline the right way to organize the reliability activities in your company to achieve reliability objectives while at the same time helping to meet time-to-market and cost objectives.

It is a mistake to copy a list of reliability tasks from some guidebook or template and think you have a decent reliability plan. It is essential to develop the appropriate set of tasks based on the unique circumstances of a given project or program and to ensure that these tasks are well written, fully understood, approved, and supported by program management and effectively executed.

1.6 Definition and scope of the reliability plan

Let's start with the technical definition of "reliability":

> Reliability is the probability that an item will perform its intended function for a designated period without failure under specified operating and environmental conditions.

The dictionary definition of reliability relates to trustworthiness. Here, it means the customer can trust your product or system to function as expected. This definition extends to brand loyalty as well. A customer may desire a product "that just works," which is not the same as the technical definition. We use the technical description as it is measurable and informative with the aim to guide the development of a product that meets the customer's expectations related to product performance over time.

A "plan" can be defined as a written account of the intended future course of action to achieve specific goals or objectives within a specific timeframe. It explains in detail what needs to be done, when, how, and by whom. Therefore, a "reliability plan" includes the entire set of tasks to achieve program reliability objectives, including responsibility for execution, timing, and resources.

The scope of a reliability plan can be as broad as a company-wide program or as narrow as a specific project. In this book, a "program" refers to the company – or division-wide business objectives, including organizational resources, and can span many projects. A "project" refers to a specific set of interrelated tasks to be executed over a fixed period and within certain cost constraints and with other limitations.

In its essence, a reliability plan is a roadmap to move a company from where it is today to where it needs to be from a reliability standpoint. It should include all the tasks needed to ensure that products and processes achieve the reliability objectives of the company. The objective is to design and manufacture a highly reliable product on time and in a cost-effective manner. The reliability plan typically includes a summarized version of the reliability strategic vision, and the full set of tasks, encompassing both reliability methods and organizational improvements, to achieve reliability objectives. This includes tasks that support each of the various stages of product development (e.g., concept, design, development, manufacturing, field use, and service), as well as tasks that support the business process (e.g., staffing, training, procedures, and resources) Everything the organization needs to do to move from the current reliability status to achieving the reliability strategic vision should be included in the reliability plan.

1.7 The art of creating a reliability plan

No plan as written is perfect since there is no way to anticipate all the information yet to be uncovered during execution of the original plan. However, you can successfully accommodate the successes and setbacks along the way and set a course to uncover what is necessary to move forward deliberately.

There is no one plan that will work within your organization in all cases. Each plan must meet the current needs given the current set of conditions and constraints. There should however be a framework to guide your plan development and adjustments along the way. What worked for your previous project will likely not work for the next one. Each and every situation and challenge is unique. Thus, every plan has to be unique.

In this book, you will not find a set of tasks to accomplish within every plan. Nor will you find a fixed list of methods and tests that will allow you to create a highly reliable product. Instead, you will find a framework with six steps to guide your thinking.

A reliability plan assists the people in your organization in making better decisions, resulting in a reliable product that meets your customers and your business objectives. A plan provides a roadmap for your team to know what the objective is and what hurdles in knowledge or understanding may prevent achieving those objectives. A well-crafted plan sets out the methods to address the creation of the specific information that enables the entire team to make appropriate decisions resulting in a reliable product.

The essence of a reliability plan is to incorporate the reliability vision, understand the gaps, and provide a robust set of tasks. Accomplishing the tasks provides unambiguous goals and objectives, adds clarity around the decision options and risks, and lays out a learning path that can be adjusted along the way based on available information. Keep in mind that a plan alone does not create a reliable product. It is the well-crafted decisions that create a reliable product. In other words, it is the thousands of decisions that occur during the creation of a product, from concept through manufacturing, shipping, and installation, that create a reliable product, not the existence of a reliability plan. Reliability is built-in primarily during the design phase by all the decisions concerning the design concept, component selection, material selection, etc. These decisions are not all laid out in the reliability plan, yet the plan can influence the decisions across an organization that impact reliability performance.

Develop Reliability Strategic Vision	Perform Reliability Gap Assessment	Identify Reliability-Related Decisions	Select the Right Reliability Methods	Create an Effective Reliability Plan	Execute Reliability Plan Tasks
1	**2**	**3**	**4**	**5**	**6**
Statement of envisioned future for company from reliability viewpoint	List of "gaps" between reliability vision and current capability	Prioritized reliability-related decisions to achieve reliability vision	Vital few reliability methods that support key decisions	Detailed reliability plan, including who, what, where, when, and how	All reliability plan tasks adjusted as needed and completed

Deliverables

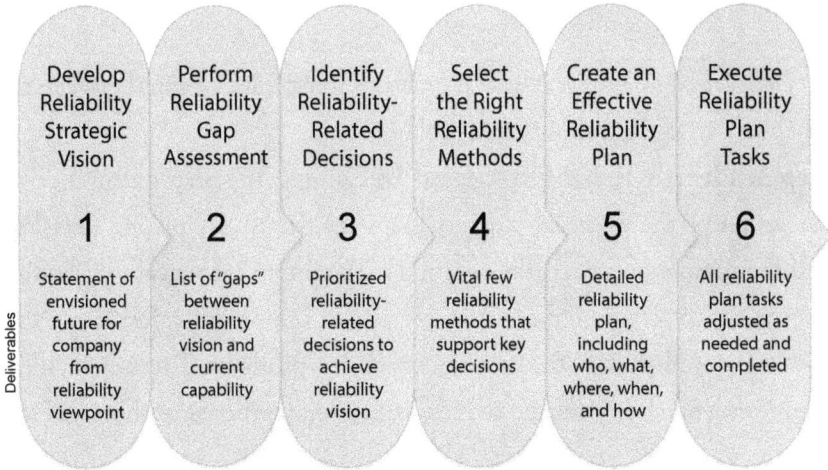

Figure 1.1. High-level primary steps to achieving high reliability.

1.8 Primary steps to achieving high reliability

Figure 1.1 Illustrates the primary steps to achieving high reliability. Here we provide a high-level summary of each step. Subsequent chapters will go into detail. See Chapter 4 for an overview of the six-step process.

Step 1: Develop a Reliability Strategic Vision. The reliability strategic vision outlines the overall vision or direction for reliability for the organization or program. It should be developed with full management support, understood by all employees, and integrated into work activities. (See Chapter 5.)

Deliverable: Statement of envisioned future for the company from a reliability viewpoint.

Step 2: Perform a Reliability Gap Assessment. A reliability gap assessment aims to identify the shortcomings in achieving the reliability objectives so that a reliability program plan can be properly developed. By definition, a reliability gap assessment is a comprehensive analysis of the specific disparities (gaps) between a company's vision for reliability and the current reliability capability, mapped to a maturity matrix. (See Chapter 6.)

Deliverable: Current reliability maturity and a list of gaps between the reliability vision and the current capability.

Step 3: Identify Reliability-related Decisions. This step narrows down the scope of the eventual plan to focus on the team's pivotal decisions and requirements concerning reliability. The intent is to create a plan that doesn't try to accommodate every request or opportunity, just those that have a major impact on the creation of reliable products. This step includes identifying reliability-related mandates and requirements, as these will also require resources. (See Chapter 7.)

Deliverable: Prioritized reliability-related decisions to achieve the reliability vision.

Step 4: Select the Right Reliability Methods. This step focuses on determining what to do to close the gaps and address key decisions identified in steps 2 and 3. By using the information from the reliability strategic vision, the results of the reliability gap assessment, and the list of key decisions, you are ready to select the right methods for inclusion in the reliability plan. (See Chapter 8.)

Deliverable: Vital few reliability methods that support key decisions.

Step 5: Create an Effective Reliability Plan. This step focuses on who should perform the selected specific activities and how and when they should be done. The methods identified in step 4 are put together in a comprehensive plan. The core of the plan is a set of tasks, including the person responsible and target completion dates. (See Chapter 9.)

Deliverable: Detailed reliability plan, including who, what, where, when, and how.

Step 6: Execute Reliability Plan Tasks. Once the reliability plan is approved by management, the implemented tasks provide value by influencing decisions concerning the achievement of the reliability vision. This includes a review of what went well and what did not. (See Chapter 10.)

Deliverable: All reliability plan tasks adjusted as needed and completed.

1.9 Value of reliability engineering and management

The primary value of product reliability is in meeting the customer's expectation that the product will work as intended over time. The market rejects products that fail too often and desires products that "just work." Creating a reliable product not only reduces warranty expenses but also saves your customers time and resources. An extension of the value consumers place on reliability is their willingness to pay a premium for highly reliable products. Automobiles, computers, printers, appliances, and test equipment are all areas where products of known high reliability can command a premium.

Paying a premium is worth it: The cost of downtime during a failure more than outweighs the additional purchase expense. In some situations, using equipment until failure is an appropriate strategy, yet that equipment has to be "reliable enough" over an expected timeframe. Products that are sought after and command a price premium lead to higher sales. Additionally, the lower failure rates reduce warranty expenses. Yes, it may cost more in materials to create a durable product. However, the increased sales and reduced warranty expenses more than offset this additional cost. The result is an increased profit margin for the company.

There are two primary ways in which reliability engineering and management create value. The first is during the product development process. An example is if potential problems are anticipated and resolved early in the product development process, this can reduce the number of problems or mistakes that occur and shorten the build, test, and fix cycles. That saves money. This is executing the plan well.

The second way to create value is by meeting customers' expectations that the product will work as intended over time and increasing the willingness of customers to buy again, tell others, or pay a premium for products with high reliability. That increases sales and market share. This is executing the right plan with the appropriate objectives.

1.10 The importance of crafting the reliability plan

Reliability plans are specific to the circumstances of the industry and company. There is no such thing as a "one size fits all" reliability plan. What works for one company in a given industry may not work for another company in a different industry.

Several factors influence the selection of reliability tasks. These factors include the type of industry, the relative cost of products, the volume of products produced, maturity of the organization, rate of technology change, and relative importance of safety and reliability to users. For example, company X designs and develops commercial aircraft. The average cost of a single product is $10 million. The company produces 100 planes per year. Its organization is moderately mature in terms of reliability capability. There is a high degree of new technology incorporated in new aircraft development. In addition, the safety and reliability implications of its product are extremely high. Compare this to company Y, a producer of retail office supplies, such as ballpoint pens. The average cost of a single product with company Y is under $1. This company produces 1 million pens per year. Its organization is new to the subject of reliability. Technology is relatively stable from year to year. The importance of safety and reliability to consumers is moderate.

The selection of reliability tasks is very different between company X and company Y. For example, company X may spend considerable time and resources using analytical techniques to model performance and reliability as part of its efforts to design-in reliability; however, company Y will probably not spend time and resources in this area. Given company X and company Y's relative difference in product cost, there will be significantly different testing and assurance task approaches. Company X has more experience and capability in the use of reliability tools than company Y, and, as a result, can take on tasks that are consistent with greater reliability maturity. The resulting reliability plans will be different.

All of this will be explained in detail in Chapters 5–10.

1.11 Reliability planning successes

As covered earlier, reliability engineering has value. For example, it can improve product reliability, increase uptime, and drive customer satisfaction. The following are stories based on real situations that resulted in significant value for the organization.

1.11.1 Reducing the expense of field failures

A telecommunications test-device company created a low-volume, very expensive test system. However, approximately 50% of the units delivered failed on installation. This resulted in the units having to be replaced, troubleshooting and diagnostic work, reworking and retesting, and significant degradation of customer satisfaction.

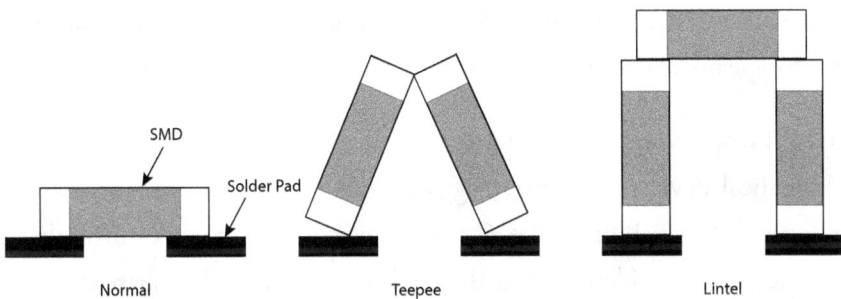

Figure 1.2. Normal surface mount device (SMD) attachment to two pads. The teepee has two components tipped upward to fit across the two pads. The lintel has three components across two pads. The solder is not shown for clarity.

One of the primary reasons for the failures was poorly designed solder joints of discrete components. Using root cause analysis, the company determined that the novel solder attachments (see Figure 1.2 for examples) were the result of the procurement manager's desire to save material costs by not buying prototype circuit boards. The alternative component-attachment techniques enabled achieving the desired capacitance or resistance without altering the printed circuit board (PCB).

Each PCB costs $50,000, and annual production for a specific product and its PCB generally was less than 50 units. With the small volume, the cost of setup dominated the cost of the boards. Consequently, buying all 50 units with one setup significantly reduced the per unit cost. The cost reduction was part of the procurement team's business objectives. Reducing PCB setup costs was an obvious action to allow the team to meet its cost reduction goal. The cost of field failures did not affect the procurement team or its cost reduction goal calculation.

The reliability engineer on the team estimated the cost of a single failure and the likelihood of component failures if they were properly mounted to the PCB and quickly determined that the savings of approximately $100,000 on materials costs resulted in approximately $10 million of annual warranty-related costs. Once this was highlighted to senior management, the procurement team began to understand the importance of prototype boards despite the increased material costs involved.

1.11.2 Improving market share

A medical device company struggled to launch a product without significant field reliability issues. Company personnel did work diligently to resolve problems identified in the field. However, market share declines were caused by the inability to meet adequate reliability expectations at product launch. The management team decided to focus on designing and building a reliable product. To do so, the team created a specific reliability plan for a new product in development. The product included new technology for the company and filled a hole in its product portfolio. It was an opportunity to both learn DFR and reestablish credibility in the market.

The specific tasks in the plan focused on identifying failure mechanisms and estimating product life based on the dominant failure mechanisms. The team set and apportioned reliability goals and worked with suppliers to source reliable parts. They designed and executed a series of accelerated life tests for areas with the highest risk. The team then identified the assembly steps with the highest risk and developed control systems to monitor and improve those processes.

A few months after the launch, when typically, many product returns would have been received, the team decided to call customers to evaluate the product's performance. There were no returns in the first three months for the new product. The customers were impressed. Market share started to increase.

As a result of this experience, the design and manufacturing teams applied the lessons learned to all the company's products. They continued to improve product field reliability and market share. Currently, the company enjoys a significant market share in a highly competitive marketplace and regularly receives feedback that the reliability of their products makes the buying choice obvious.

1.11.3 The 10× example

In the 1980s, Hewlett-Packard (HP) embarked on a corporate-wide program to reduce warranty expenses by 10× in less than 10 years. A team surveyed the various product divisions halfway through the decade and found a relationship between active and regular use of key DFR activities and operating profit (Moss 1996, p. 5.4). Figure 1.3 shows the relationship between the percentage of use (average use of eight DFR tools) during development projects and the corresponding pre-tax operating profit. The

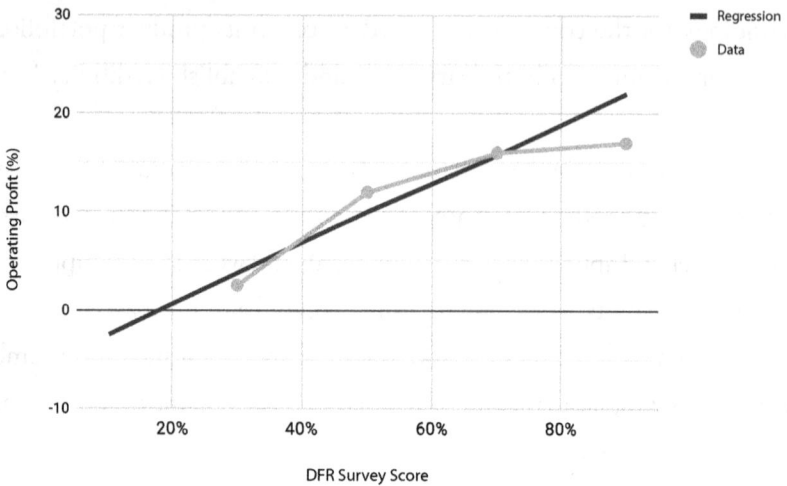

Figure 1.3. Plot of operating profit versus DFR survey score.

relationship illustrates the benefit to profit through the active application of DFR practices within an organization.

The improvement in HP's product reliability reduced warranty and related expenses. Note that warranty expenses, which include call center, shipping, repair, or replacement expenses, comprise only a fraction of the total cost of product failures. Beyond those common warranty expenses, the organization may incur additional costs of redesign and testing time, scrapping inventory, and reconfiguring supply chains and product systems.

To appreciate the magnitude of warranty expenses, one can review warranty data as reported by public companies in the United States since 2003. Eric Arnum tracks and reports on warranty accruals and expenses in the online newsletter, Warranty Week. Figure 1.4 shows a plot of warranty claims paid worldwide by U.S.-based companies. Warranty payments in 2019 were approximately $25.9 billion (Arnum 2020). See WarrantyWeek. com for additional warranty data, trends, and analysis.

The 10× push at HP occurred during the 1980s and many companies at the time treated warranty expenses as a trade secret, yet internally the CEO

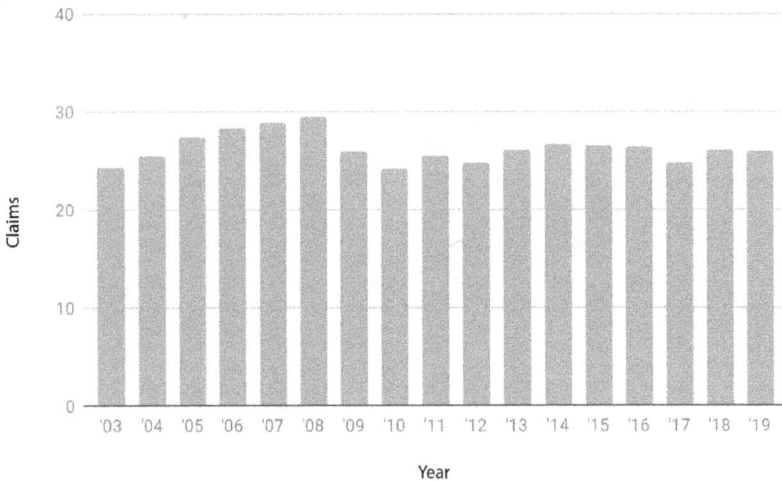

Figure 1.4. Worldwide warranty claims in billions of U.S. dollars.

realized the large cost of unreliability and decided to focus on improving reliability as a result. Reducing warranty expenses was a driving factor in HP's 10× program. The company also realized additional value with reduced development costs, premium pricing, and increased market share.

1.12 Summary

To meet the reliability challenge, we recommend creating and executing a reliability plan that is crafted to the unique circumstances of your organization. To do so, we are introducing a six-step process that leads to the creation of a highly reliable product.

In the next chapter, we will explore the different approaches to reliability management and some fundamental reliability concepts to provide a foundation for implementing the six-step process to create a highly reliable product.

Chapter 2

THE ESSENCE OF RELIABILITY MANAGEMENT

In matters of style, swim with the current;
in matters of principle, stand like a rock.
Thomas Jefferson

In this chapter. We discuss a few foundational concepts that shape how to think about and craft a reliability plan. Topics include the two different approaches to creating a reliable product, the necessary focus on designing for reliability, and the information provided by failures. Also provided is a brief introduction to understanding how reliability activities must generate value and what we mean by value.

2.1 Quality and reliability cannot be delegated

W. Edwards Deming (2000, p. 21) has provided guidance on the subject of delegation, stating that "It is not enough that top management commits themselves for life to quality and productivity. They must know what it is that they are committed to—that is, what they must do. These obligations cannot be delegated."

A common expectation consumers have concerning products is that the product works. The product provides value by performing one or more functions. In addition, implicit with this expectation is that the product will function over some duration. A reliable product meets or exceeds this common expectation. Every product has a finite duration of successful operation before failure occurs. Matching the products' capability to last sufficiently long to meet expectations is one of many tasks facing product developers. Every product has a reliability attribute—whether or not it is known. The management team determines the desired reliability and the means to create a product that meets the reliability objective.

Reliability management includes elements of leadership as well as the alignment of resources to achieve business objectives. Business leaders set the organization's pace, tone, priorities, and decision criteria. This is done by actions managers take on a day-to-day basis and based on policy guidance documents. The leadership team commonly establishes a product development life-cycle document that includes stages with reviews. By experience, life-cycle documents often do not include reliability directives explicitly. The better life-cycle documents include guidance to frame reliability, objectives, risks, and priorities.

Senior management establishes the priority and importance of product reliability. It is not possible to delegate this leadership role. Product development team members have to make decisions that will impact product reliability. If there is a lack of leadership, decisions may not include reliability considerations or the decisions may over prioritize the importance of reliability. Left without reliability guidance, the product's design team may or may not meet the customer's reliability expectations or business objectives.

Consider the following example:
Shortly after the purchase of the Enthusiast Series, you (the reader) sit down with two development engineers and their manager. The discussion turns to reliability goals for their current project.

You ask the first engineer whether he knew of a reliability goal. "Yes, it is 5,000 hours MTBF [mean time between failures]," he replies quickly. You ask what that goal means and how it influences his design work. He replies that this product's goal is very difficult to achieve. He talks about needing expensive parts, the best vendors, and careful testing.

The second engineer also knew the goal. When queried about how that influences his work, he replies that that goal was easy to achieve. He uses the least expensive parts (any vendor's parts would work), and there was little need for testing other than basic functionality. At this point, the interviews became more interesting, as both engineers worked on the same product.

The team manager also knew the reliability goal. She said that it was a meaningless goal because they don't measure it at any point during the development process.

All three worked on the same product and had different reactions to the knowledge of the reliability goal. One of the roles of a leader is to establish expectations. A reliability goal for a product is one such expectation. By not also measuring the reliability performance, the goal alone led to competing activities within the team and a lack of the team's ability to achieve the stated goal.

A key role of management is to reinforce and support the expectations concerning reliability. This cannot be delegated. The behavior and actions of management establish the culture concerning reliability.

Managers also are involved with the allocation and alignment of resources. In product development teams, these resources include design talent and money. Money is often the common denominator across an organization and thus is the language of management. Talking about reliability in terms of money, profit, or warranty lets the management team understand a reliable product's relative importance and value.

The other primary factors of cost and time to market often translate directly into money. Achieving product reliability requires an investment to avoid future product failures and associated costs. The team making the investment may not directly feel the impact of future product failures, but this aspect cannot be neglected.

2.2 Fundamental philosophies for creating a reliable product

There are two fundamental philosophies to consider when addressing product reliability: reactive and proactive. Every organization must have the ability to react to failures yet avoiding or minimizing failures is optional and preferred. The reactive approach has the team respond to failures as they occur during prototyping, production, and customer use. The proactive approach focuses on preventing issues in the product design, assembly, and use before they can lead to a product failure. Thus, the preferred approach is the proactive one.

2.2.1 Reactive approach

The reactive approach focuses on identifying failures, conducting root cause analysis, and implementing the appropriate corrective action. This may include a redesign of the product or a change in suppliers. Designers often believe that having a concrete product failure is essential to solving reliability issues. This approach does have the limitation of waiting until something goes wrong before acting.

The reactive management style is characterized by senior management paying attention to product reliability only when there is a major field failure problem. The basic message from management is to focus on time to market, bill of material cost, or product functions and not reliability. In some cases, the individuals who are particularly adept at solving field failure issues are rewarded and considered heroes. This culture quickly devolves to one in which field issues regularly occur, pulling engineering talent from new product design tasks.

A culture of responding to reliability failures may require significant engineering resources to address field issues. This culture may also impact product development in an adverse way (Repenning et al. 2001). Our experience with reliability program assessments reveals that an estimated 25% of engineering talent is committed to addressing design and manufacturing errors related to reliability. Some organizations create entire departments, e.g., sustaining engineering, to "shield" the product design teams from re-engineering products failing in the field.

2.2.2 Proactive approach

In contrast to the reactive approach, the proactive approach focuses on preventing failures before they occur. No design is perfect, yet some are better than others, especially concerning product returns or lost sales. A product that does what it is expected to do and does it over time is considered reliable. For example, the HP calculator had that reputation and enjoyed an associated very low field failure rate.

The investment into creating a robust and durable design requires informed design practices, the anticipation of failure modes and mechanisms, and resources to enable essential design evaluations and improvements. The combination of setting a goal and measuring progress toward that goal enables the design team to create a reliable product. Measuring reliability, both in terms of failure rate and money, enables the entire design team to make informed decisions that consider the impact of cost, time to market, and reliability.

In the proactive approach, a variety of tools are used prior to failures to better understand the product design. This helps avoid a wide range of possible failure mechanisms. This methodology requires understanding materials, design performance, product use, and environmental stresses.

In the proactive DFR approach, with an understanding of the full range of stresses and incidents that may occur to the product, the design team is then able to design a product that will continue to function within the expected environment. Although it may not be feasible to create a product that will work across every set of possible stresses the future individual products may experience, making informed decisions that minimize the field failure rate and customer experience is possible.

Product reliability occurs due to how well the product performs over time within specified environmental and use conditions. Reliability performance results from decisions and actions during design, assembly, and use. A well-designed product will meet or exceed reliability expectations. Likewise, a weakly designed product will experience more failures than expected. Well-crafted design decisions that accommodate all expected use conditions determine the resulting product reliability. This is especially true given that the added variability and errors in an assembly process will only make resulting products less reliable.

2.2.3 Reliability maturity

Maturity refers to the behaviors within an organization. A mature company is able to repeatedly create reliable products. An immature company's erratic processes may or may not create a reliable product. Maturity reflects the culture or approach to reliability. Immature organizations tend to ignore reliability or use crude techniques to set requirements, identify risks, or measure results. Mature organizations proactively work across the organization to enable appropriate decisions using the best techniques fit for the task.

The reliability maturity matrix is a framework to help you understand the organization's reliability culture and to make improvements to that culture, as needed. See Chapter 3 for additional discussion of reliability maturity and Chapter 6 for application of the maturity matrix to the bicycle examples.

In any organization there is a mix of proactive and reactive behavior. The management team, in large part, sets the norms and culture surrounding the organization's approach to reliability. Given the significant benefits to your organization and customers, the actions, rewards systems, and priorities should align toward working to prevent failures in a proactive manner. Of course, failures and surprises will still occur. The organization needs to respond and correct failures diligently, yet not to the exclusion of preventing failures from occurring. When proactively designing in reliability, there will be fewer surprises and critical failures that require immediate attention.

2.3 Design for reliability

There are two basic approaches to designing a reliable product: 1. build, test, and fix and 2. analytical (design reliability in). Both approaches can lead to the creation of a reliable product. Both have benefits and weaknesses. Both are employed to some degree in many cases as they are to a great extent complementary. The actual reliability performance of a product is predominantly determined by the decisions made during the design and development of the product.

The build, test, and fix approach is best illustrated by a rapid prototyping development process. As soon as possible and as often as possible, you should build prototypes of the product, evaluate and test the prototype to find flaws and weaknesses with the design, and then redesign and build more prototypes. An underlying premise is that any design has a finite number of flaws and the ability to rapidly identify and resolve those issues helps you to quickly develop a reliable product. This approach has the benefit of identifying and enabling characterization of design flaws, which, in turn, enables the design team to remove those flaws before they lead to product field failures. This approach works well with products that lend themselves to rapid prototyping and testing.

The analytical approach is characterized by the use of DFR tools and modeling of failure mechanisms. The underlying idea entails understanding and characterizing the possible failure mechanisms so that the team can estimate the potential failures resulting from a wide range of varying stresses. In systems with a few dominant failure mechanisms that have well-defined models relating stress to degradation or probability of failure, this analytical approach works well.

The analytical approach includes thermal modeling, stress–strength analysis, and physics-of-failure modeling, for example. The premise is that everything fails, and understanding the relationship between stress and failure enables estimating the time to failure. This also works especially well for accumulated damage or wearout-type failure mechanisms. The analytical approach is useful for expensive systems for which it is cost-prohibitive to build many full system prototypes or when testing the system is not practical. This approach does rely on previous work to allow you to fully understand the relationship between stress and failure.

Both approaches have the benefit of focusing the team on failure mechanisms. Both have the advantage of returning valuable information back to the design team, enabling product-reliability-influenced decisions.

The benefits are the following:

In the build, test, and fix approach:
- Tangible failures enable detailed failure analysis.
- System interactions are included.
- There is an increased possibility of finding previously unknown failure mechanisms.

Meanwhile, the analytical approach
- is not limited by sample restrictions (i.e., limited number, not final construction methods or materials),
- can cover the range of stress values on failure rate impact,
- enables what-if analysis very early in the design, and
- can quickly yield results compared to testing times (with the right models).

However, there are limitations:

The build, test, and fix approach
- requires the ability to create prototype units quickly,
- tends to suboptimize a robust solution, and
- can take a long time if tests are not properly accelerated.

Meanwhile, the analytical approach
- requires analytical capability,
- takes an investment in creating characterization models, and
- tends to focus on known failure mechanisms.

An essential element of DFR is finding failure mechanisms and eliminating or minimizing them during the design process. Some tools support discovering failure mechanisms, allowing the team to take action. Other tools support understanding failure mechanisms, again allowing the team to take action.

The decisions during the design process establish a product's reliability. Both approaches provide information to the decision makers that enable them to fully consider the impact of their decisions on reliability performance. The effectiveness of both approaches relies on creating useful insights to fully inform decision makers. Most projects benefit from a hybrid approach, combining analytical methods, where possible, along with product assurance, such as test and fix.

2.4 Understanding how and when items fail

A core element of designing a reliable product is the acceptance that every product will fail at some point. Each design decision incorporates the desire that the product should survive some set of conditions over a finite period. Central to supporting these decisions is helping your team understand where and to what extent the risk of failure exists. For the identified failure risks, the question quickly turns to "When will the failures occur?"

Engineers and business leaders tend to expect products to work. They think in terms of a "success frame of reference." Reliability professionals, in contrast, tend to consider what and when items will fail or take a "failure frame of reference." By highlighting the value of identifying and understanding failures, we can shift the organization's culture to one that celebrates failure by appreciating the insights and value each failure provides.

2.4.1 What will fail?

Design engineers naturally consider the potential use of a device and the likely weaknesses of the design (Petroski 1994). The intent is to provide the desired functions without failure. Even for relatively simple products, hundreds of potential failure mechanisms compete to cause the device to fail.

Reliability engineering toos of risk prioritization and discovery can be combined with engineering knowledge to identify the most serious and likely failure mechanisms. An organization's reliability culture should encourage and reward the finding of failures. Only when armed with knowledge of failures can the design team decide to make improvements. Understanding the risk of failure informs decision makers as they allocate resources appropriately.

Product testing should focus on testing to failure and thorough failure analysis. The detailed information on the root cause of failures provides the necessary information to consider appropriate remedies. Without identifying potential or revealing actual failures, the design team is left to speculate as to which design elements may require improvement. In some cases, modeling and simulation replace prototype testing, which may be supplemented with material – and component-level reliability characterization.

The awareness of what may fail, along with tangible evidence of what actually fails, starts the discussion on reliability improvements. The reliability project supports creating detailed knowledge of the failure mechanisms with infrastructure (tools and methods). The work to find failures requires a culture in which finding failures is a desired and celebrated activity.

2.4.2 When will it fail?

Knowing all the ways an item may fail may provide an overwhelming set of opportunities to improve the design's reliability performance. One key element of information useful for narrowing down the options for improvement is the expected time before the failure mechanism will lead to field failure. A faulty assembly or inadequate design may lead to early life failures in every device or a selected component may degrade in performance slowly, leading to failure after many years of operation.

Knowing how and when a failure will occur provides crucial information enabling prioritization of which failure mechanisms require attention. Some may require changes in materials or components, some may require adjustments to the assembly process, and others may require addressing diagnostic and repair capabilities.

A reliability project should support the exploration of the failure mechanisms' time-to-failure behavior. This includes modeling, simulation, environmental and use characterization, along with ALT for materials, components, and assemblies. The program infrastructure includes a mix of modeling and laboratory resources, enabling the team to estimate the time-to-failure distribution for the most likely failure mechanisms. The reliability work reduces uncertainty concerning when failures may occur or under what scenarios failures may occur.

2.4.3 Risk of failure

Not all failures are the same. Some cause catastrophic damage, while others lead to no perceptible change in product performance. The timing of when a failure occurs may also change the resulting risk of the failure. Early failures often damage brand reception and may lead to reduced market acceptance.

Failure during the warranty period increases financial obligations. In any case, failures at any time increase the cost of ownership to the customer.

Ideally, the reliability program includes the infrastructure and support to estimate the total cost of a failure. The cost information should range from the direct cost of the failure (e.g., time to reset the system, component replacement, or system replacement) to the impact on the customer (e.g., lost productivity) and the impact on customer satisfaction. In some markets, the increased risk of liability or recalls is also a factor.

The reliability project enables you to estimate the risks in the appropriate context with the correct perspective. If there is a finite chance of an expensive failure, the information should be clear and useful for the decision makers to evaluate.

2.4.4 Celebrating failure

It is the combined knowledge of failure mechanisms, time-to-failure probabilities, and financial impact estimates that create clear information for decision makers across the organization. Knowing only one of these elements is insufficient to properly balance the risks and options to make the right decision. Each product and market is different, yet the ability to make informed decisions considering the reliability performance of your product is important, regardless of the product or market. The reliability tools, methods, and culture employed are a direct result of the reliability program.

Keep in mind that each portion of your program has a role to play in addressing one or more of the three elements discussed above. When all three elements are regularly discussed and addressed by teams across the organization, then your program will be successful.

Celebrating failures means shifting the focus to finding and understanding potential failures.

2.5 Calculating value

There are two common ways to calculate or estimate the value of a reliability activity. The first is by executing a reliability method and tracking the impact it makes and directly measuring the difference that method created. The second is by not waiting till the results are measurable, such as when making a proposal. In this case, we have to rely on detailing the logical connection between what the expected reliability method's output creates or provides and how that output leads to the creation of value.

By definition, "value" is

> the regard that something is held to deserve; the importance, worth, or usefulness of something (McKean 2005).

As reliability engineers, we work across the organization to bring a reliable product to market. The value of meeting the customer's reliability expectations results in customer satisfaction, increased sales, and in some cases premium pricing. One dilemma to avoid is only reacting and solving field problems instead of identifying and avoiding those issues in the first place. Some organizations reward "firefighters" yet the real value to the organization is in avoiding the same issues entirely (Repenning and Sterman 2001).

We want a reliable product.

Being a pivotal element in the process means you have provided value to the organization and to its customers. Adding value increases your opportunities for career success. Even if the product does not succeed in the market, adding value to the program increases your chance of career success.

In the business world, value is money.

If the return on investment (ROI) is adequate in comparison to other potential investments, then it is acceptable to make the investment. For example, if the cost to conduct an accelerated life test is $50,000, will the value of the test results be worth at least ten times the cost? If not, then the expected value of the test results does not justify conducting the test. Each organization may have a policy on what constitutes an acceptable ROI.

If the early prototype highly accelerated life testing (HALT) reveals three critical design faults, this provides the design team with time to resolve the issues without delaying the product launch. The delay may cost lost sales and depends on your market.

In each case, the reliability engineer's task is to recommend and execute tasks that affect decisions, reduce risk, save time, or add value. This extends to every encounter with your fellow engineers and managers working to bring a product to market. You can provide insight, information, and knowledge that enhance the entire team's ability to create a reliable product.

Adding value should be a habit.

For the larger tasks that require significant resources to accomplish, you may have to estimate the ROI before being provided with the prototypes and equipment to accomplish the task. In other cases, before starting a task, you may need to determine how and where the resulting information will be used.

It is crucial to meet key deadlines because even perfect information for a key decision a day late is not useful. For each activity you start or recommend, you need to understand the cost and return to calculate the ROI; if the activity does not have value, it is time to focus on something that does. Having a habit of adding value, while being able to articulate the value you have contributed, lets you clearly focus on activities that provide the greatest benefit to you and your organization.

Here is a summary of six ways you can find value from reliability tasks:

Cost of warranty reduction. Has the task directly identified or mitigated a potential field failure problem? If so, estimate the probable cost of the field problem and how the work reduced the probability of the field failure occurring. Has the task provided knowledge or a process that will reduce the risk of field failures in other products?

Time-to-market impact. Did the task help your organization meet or beat your time-to-market target? Did the task serve to identify any problems that would have delayed your project? If so, estimate the duration of the delay along with the expected cost of the delay.

Time-to-volume impact. Did the task help to accelerate or meet your time-to-volume (production ramp schedule) goals? If so, estimate the reduction in expenses or costs by avoiding the problems identified.

Material costs. Did the task avoid or save on any direct product material or test equipment costs?

Customer satisfaction. Did the task contribute to improving customer satisfaction with the product? If so, estimate how a reduction in customer calls to a call center or requests for assistance impacts cost. If you have a model of the expected additional sales and customer satisfaction (check with marketing), estimate the increase in revenue resulting from improvement in customer satisfaction.

Opportunity cost. Did the task reduce the time the engineering staff spent resolving reliability problems? If so, estimate the value of the engineering work they are now free to complete.

In summary, it takes practice, yet you will naturally look for and quantify value for each of your reliability-related activities over time. As you do so, you will be able to clearly estimate the ROI for proposed tasks and track and qualify the value created as a result of specific tasks. In large part, it's all about learning to talk as your management team does about investments, opportunities, and profits. Doing so increases your influence within an organization and improves product reliability for your customers.

In some circumstances, warranty and warranty cost reduction are not always the most important value that we can provide. There may be circumstances in which warranty cost was already relatively low, and reducing time-to-market, material costs, or call-center costs may have a more meaningful impact on organizational profit. See the book Finding Value (Schenkelberg 2014) for more information on articulating reliability-task-initiated value creation.

2.6 Keeping it simple

A reliability plan is a means to the end of a reliable product. It is not a means in itself. Simply listing a dozen or so reliability engineering activities and checking them off once accomplished may not achieve the desired outcome. A simple reliability plan that achieves the desired outcomes takes work to achieve. It takes understanding the current situation and information available for a program. Adjusting the tasks to fit the risks and needs of a program is essential and adds value by informing decision makers.

Focus on what is important. This is often called the Pareto principle or the 80/20 rule. The Pareto principle states that roughly 80% of consequences for many outcomes come from 20% of causes (the "vital few"). Build the reliability plan to address the areas of highest risk to the creation of a reliable system. The risk of failures described in Section 2.4.3 is one factor for prioritization. Another is how the results of the task will impact critical design and development decisions.

Consider this example:

In the Intro Series bicycle line, a senior manager learned that HALT would be useful and gave it a try. The team learned about a couple of new failure modes for a new component, which saved them the anticipated warranty expenses of field failures. Therefore, this manager mandated that all programs use HALT on every prototype during development.

The intention may have been good. In practice, the organization learned that sometimes HALT is an appropriate tool. They learned that conducting HALT on low-risk designs or on a product with a long list of known issues would yield very little new information. The HALT exercise would often yield useful information for products with new materials, new design elements, or new vendors. Over the period of a year, the team added a decision point to the process: Do we need to do HALT on this prototype?

It's the same for each element of the program. The specific task has to be useful. That is one way to filter down what is possible to what is necessary.

Another consideration is cost. Each reliability activity has a cost. It may be engineering time or expenses for the prototype and testing equipment needed to accomplish the activity. We are almost always limited by budget, either by time or money. By creating a plan that addresses the areas of highest risk to meeting business objectives related to reliability, we can narrow down and simplify the reliability program.

One approach when limited by budget and still wanting to accomplish a wide range of reliability and environmental testing is to minimize the chance the testing will damage the prototype. Maybe we only have five units. By keeping the testing stress to a minimum, all units can survive the testing. This may include 20 different tests for each unit.

Using this approach of moderate stress across a wide range of tests achieves very little. Rather, with the earlier work on risk and discovery, along with engineering judgment, we can focus the use of the limited number of units to those areas that may reveal the most useful information for product reliability improvement.

Now consider the next example:

The Pro Series previous development program had 35 different reliability tests. To accomplish all the testing, each test was minimized in terms of duration, stress, and number of samples. It was a logistical challenge to accomplish all the testing in the time between prototypes becoming available and the design review meetings.

As field issues arose, the organization's response was to add another test that would catch that defect, thus preventing it from becoming a field issue in the future. The list of such tests now grew to 36. Two things resulted: 1. New issues continued to occur with field products and 2. some of the same issues arose, despite units passing all the tests.

The ability of the testing to reveal new or known issues is minimized by the inability of the specific tests to accomplish their purpose. The Pro Series team decided to only conduct testing to answer a question essential to the current project risks and uncertainties.

Reliability engineering is not limited to conducting testing. Reliability engineering is the use of information to improve the design and production of products such that they meet customer reliability expectations. In some cases, reliability testing is the right tool to create results that inform the decision-making process during design.

2.7 Integrating with product development

Achieving reliability is a team activity. Reliability objectives cannot be achieved by the reliability engineer alone. The best way is to enable individual design and process engineers to make sound and effective reliability decisions.

Engineers, managers, and vendors have entire teams working and making decisions to create a product. Since reliability occurs at the point of decision, the role of the reliability professional is to enable all of them to make great decisions. They need data, information, insights, and direction. They need the right information at the right time.

A reliability program is not something that happens after a design is finished. Reliability occurs during the design process at the point of each decision. From selecting the architecture to materials to specific components, each decision has an impact on the final reliability performance. Here is another example:

As the news broke that the company planned to purchase the Intro Series bicycle line, an engineering friend calls with the story of his brief interview with the engineering manager. Shortly after the interview started, he was asked how he would help to improve the reliability of the upcoming product. He mentioned that he would work with the development team of engineers to understand the risks, identify reliability issues, and improve the design to achieve the desired reliability. Part way into the response, the hiring manager interrupted him to say, "You will not be allowed to work with the development team."

When he realized that the manager was serious that the role of reliability engineer would not include the ability to influence the design, the caller thanked the engineering manager for her time and wished her luck with the new product line. Then he got up and left.

It is not possible to test-in reliability, nor is it a viable solution to quickly identify field issues and fix them as production continues. The design has to be right for a high-volume consumer product from the start.

One of the issues reliability engineers face with any design team is credibility. We are generally not highly skilled mechanical, software, electrical, or industrial design engineers but we have to work in these areas to make improvements to product reliability. Moreover, we often focus on areas of a design that will likely fail. We can be seen as only bringing bad news to the design team, offering criticism of the design and, to some extent, the design team's ability. We bring potential and real faults to light. As one designer said during a short meeting covering an initial assessment of his design, "When are you going to bring solutions, not just problems?"

In one case, early in a development program, the design included an edge connection for a circuit board. The board would be vertical inside the product, only supported along a short edge. In the product operation, a regular lateral motion could potentially cause the circuit board to flex. Bending a circuit board tends to cause cracking of traces, vias, and solder joints. Circuit boards, when unreinforced, are not rigid enough to prevent bending.

The mechanical engineer on the design team explained the complexities and cost of adding support for the circuit board that would address the concern for bending-induced damage. She basically said it was not in the plan and would be expensive to implement.

It happened that the product testing lab had just bought a high-speed camera. To investigate the problem, a working prototype with the vertical edge supported board was brought to the lab to determine the amount and nature of the board bending. Shortly after starting the camera and powering on the product, the product stopped working in just a few seconds.

Review of the slow-motion clip made it obvious why the system failed: A capacitor from the midsection of the board sailed off the board! The bending motion damaged the capacitor solder joints and flung the part away from the board. Without narration, that short film convinced the engineer to attach and support the circuit board to prevent board flex.

As a result of this investigation, the next set of recommended improvements for product reliability was not quickly dismissed. Building on the success and highlighting solutions to reliability issues earns you credibility. It takes time, and it will take work.

Product development starts with the initial idea and continues through the entire product life cycle. Decisions about functions, intended environment, and use, along with expected useful lifetime, include decisions that impact the product's reliability. Vendor and material selection, assembly technique design, product design improvements, adjustments required because of supply chain changes, and customer feedback all include decisions that will change the product reliability performance.

A primary role of a reliability professional is to integrate the information necessary for reliability decisions at each step of the product life cycle. As we have stated, product reliability occurs at the point of decision. When an engineer selects steel over a polymer for a structural element of a bicycle, this decision is based on mechanical, aesthetic, cost, and production considerations, as well as customer expectations, environmental use, and durability. Decisions made during product development are about finding the balance between many competing goals to identify the solution for a particular product.

Reliability is one of the considerations, but it is often difficult to measure, articulate, and understand without supporting information. The relatively simple task of selecting a material requires knowledge of a wide range of potential failure mechanisms and how they respond to environmental and use stresses. The decisions might involve not only the immediate strength of the material to maximum loads but may require information about wear and aging properties over time.

For example, if a change from one material to another would save $10 on the cost of a bicycle, but shortens the expected useful life by one year, is that a fair or reasonable tradeoff? The team will require additional information to answer this question, including a clear reliability goal that reflects customer and business expectations. It requires information on failure mechanisms, the cost of failures, the nature of the failure (catastrophic or degradation-of-performance-type failures), and information about the relationship to other elements of the bicycle's overall reliability performance.

Management of the information related to reliability, because it is connected to aspects of product development and life cycle, makes reliability engineering an important and meaningful function within an organization.

2.8 Matching the plan to culture and capability

The reliability plan has the role of guiding the team as they work to achieve the reliability-related objectives. Each step and milestone within the plan is essentially a goal in itself. As with goal setting in general, the tasks within a plan should be specific, measurable, achievable, realistic, and time-related (Doran 1981). These tasks can be described as follows:

Specific. The statement of the task within a plan is not just "do a design FMEA," for example. Rather, it specifically stipulates initiating a Design FMEA on item ABC, using team DEF, beginning by date GHI, completing by Date JRK, and using procedure MNO. This will help the team discover the important potential failure modes and prioritize actions to design out or mitigate the identified failure modes. Specific means that the goal for the task answers the who, what, where, when, and why questions.

Measurable. A goal or task has an outcome and some criteria for success. Try to avoid binary measures such as just answering whether the FMEA meeting occurred. Instead, in the case of an FMEA, use a measure that focuses on the resulting benefits, such as the number and value of actions taken based on the FMEA study.

Achievable. The implementation of the task has to be something the organization has the capability to accomplish. Including a task that requires advanced statistical knowledge such as a complex designed experiment to a team that at best calculates averages will require additional tasks and time to build the capability first. Include tasks and recommendations that build on the existing capability to step-by-step improve the organization's capability.

Realistic. The task does not stand alone and is an activity that moves the project forward toward achieving the reliability objectives. It also should be a task that fits within the overall plan with meaningful connections and relationships with the overall project and other reliability plan tasks.

Time-related. Include a deadline or time frame. Select tasks that can be accomplished given the time constraints within the project. For example, recommending an accelerated life test for a new bicycle frame material that will provide results six months after the final design review will not influence the decision of whether or not to use the new material. Instead, align the timing of tasks to provide results useful for decisions when the decisions are being made.

You can best improve the usefulness of the reliability plan by matching or only slightly altering the overall culture of the organization. Recognize that any task that challenges the cultural norms may require additional justification and reinforcement of the provided value of the specific task. For example, if the existing culture tends to promote quickly assessing problems and jumping to conclusions and implementing fixes, and a recommended task is to conduct in-depth root cause analysis, this presents a cultural conflict. In this example, making the case for a root cause analysis will provide sufficient information to truly solve the problem and avoid the expense of attempting to "fix" the issue multiple times.

2.9 Beginning with the end in mind

Business strategy, program goals, project plans, and specific tasks each describe an end state or desired outcome. The business strategy may include a vision to guide the long-term investments in product development. The reliability goal likewise defines the desired ability of a product to perform without failure for the customer. Individual tasks within a reliability plan define the tools and techniques necessary to discover, learn, characterize, predict, or understand some specific aspect related to the overall project reliability goal. If the plan includes a set of environmental stress tests, the plan should include enough detail to fully interpret the results.

For example, should testing three handlebar mounts for 100 hours in a thermal chamber produce a meaningful conclusion? What is the test or any task supposed to accomplish? What are the criteria? How will the results improve decision-making in some tangible manner?

State the goal of each task in the plan. Include how the expected results connect to other tasks and decisions within the overall project. By stating the expected results, the team will know when the task is accomplished to the desired standard, and the team will incorporate the task results into the development process, improving their decision-making.

Being clear about the results of the tasks is our preparation to recognize deviations from what we are looking for. Setting expectations enables those involved with the task to recognize results that do not meet the expected results. This recognition of anomalous results may lead to follow-up investigation to understand the reason for the unexpected results. Think of each test in particular as a hypothesis test. Do the results provide evidence that you have not met your expected conclusions? If so, it's time to explore why, adjust or improve the design, or alter expectations going forward.

2.10 Summary

The blend of how an organization approach's reliability, treats failures, and works to design-in reliability reflects the set of behaviors, norms, and leadership present within an organization. How an organization addresses reliability reflects that organization's culture.

In the next chapter, we will explore a tool that is useful for identifying the maturity level of the culture surrounding reliability.

Chapter 3

RELIABILITY MATURITY MATRIX

Wisdom is really about figuring the long-term consequences of your actions.
Naval Ravikant

In this chapter. We introduce the reliability maturity matrix, a tool to assist in understanding an organization's culture related to decisions resulting in the current or desired product reliability performance. The matrix has 11 categories (rows) that summarize behaviors for each of the 5 stages of maturity (columns).

3.1 What is a reliability maturity matrix?

A reliability maturity matrix is a tool for describing a company's approach to achieving product reliability performance. It is based on observing how an organization uses reliability methods to make decisions that impact reliability performance. A reliability maturity matrix provides a useful way to identify strengths and weaknesses, while illustrating improvements to policies, processes, and decision-making that improve the organization's ability to create highly reliable products consistently.

The matrix is organized with 11 measurement categories (rows) and with four groups of categories: requirements, engineering, feedback

Reliability

		Stage 1: Uncertainty	Stage 2: Awakening
Requirements	Requirements & Planning	Informal or nonexistent.	Basic requirements based on customer requirements or standards. Plans have required activities.
	Training & Development	Informally available to some, if requested.	Select individuals trained in concepts and data analysis. Available training for design engineers.
Engineering	Analysis	Nonexistent or based on manufacturing issues.	Point estimates and reliance on handbook parts count methods. Basic identification and listing of failure modes and impact.
	Testing	Primarily functional.	Generic test plan exists with reliability testing only to meet customer or standards specifications.
	Supply Chain Management	Selection based on function & price.	Approved vendor list maintained. Audits based on issues and critical parts. Qualification primarily based on datasheets.
Feedback Process	Failure Data Tracking & Analysis	May address function testing failures.	Pareto analysis of field return and internal testing. Failure analysis relies on vendor support.
	Validation & Verification	Informal and based on individuals rather than process.	Basic verification that plans are followed. Field failure data regularly reported.
	Improvement	Nonexistent or informal.	Design & process change processes followed, corrective action taken.
Management	Understanding & Attitude	No comprehension of reliability as a management tool. Tendency to blame engineering for 'reliability problems.'	Recognizing that reliability management may be of value but not willing to provide money or time to make it happen.
	Status	Reliability is hidden in manufacturing or engineering departments. Reliability testing probably not done. Emphasis on product functionality.	A stronger reliability leader appointed, yet main emphasis is still on an audit of initial product functionality. Reliability testing still not performed.
	Cost of Unreliability	Not done other than anecdotally.	Direct warranty expenses only.
	Prevailing Sentiment	"We don't know why we have problems with reliability."	"Is it absolutely necessary to always have problems with reliability?"

Maturity Matrix

Stage 3: Enlightenment	Stage 4: Wisdom	Stage 5: Certainty
Requirements include environment and use profiles. Some apportionment done. Plans have more details with regular reviews.	Plans are tailored for each project and projected risks. Use of distributions for environmental and use conditions.	Contingency planning occurs. Decisions based on business or market considerations. Part of strategic business plan.
Training for engineering community on key reliability-related processes. Managers trained on reliability and lifecycle impact.	Reliability and statistics courses tailored for design and manufacturing engineers. Senior managers trained on reliability impact on business.	New technologies and reliability tools tracked and training adjusted to accommodate. Reliability training supported.
Formal use of FMEA. Field data analysis of similar products used to adjust predictions. Design changes cause reevaluation of product reliability.	Predictions are expressed as distributions and include confidence limits. Environmental and use conditions used for simulation and testing.	Lifecycle cost considered during design. Stress and damage models created and used. Extensive risk analysis performed as needed.
Detailed reliability test plan with sample size and confidence limits. Results used for design changes and vendor evaluations.	Accelerated tests and supporting models used. Testing to failure or destruct limits conducted.	Test results used to update component stress and damage models. New technologies characterized.
Assessments and audit results used to update AVL. Field data collected and failure analysis performed related to specific vendors.	Vendor selection includes analysis of vendor's reliability data. Suppliers conduct assessments and audit of their suppliers.	Changes in environment, use profile, or design trigger vendor reliability assessment. Part parameters and reliability monitored for stability.
Root cause analysis used to update AVL and prediction models. Summary of analysis results disseminated.	Focus is on failure mechanisms. Failure distribution models updated based on failure data.	Customer satisfaction relationship to failures is understood. Prognostic tools used to forestall failure.
Supplier agreements around reliability monitored. Failure modes regularly monitored.	Internal reviews of processes and tools. Failure mechanisms monitored and used to update models and test methods.	Reliability predictions match observed field reliability.
Effectiveness of corrective actions tracked over time. Identified failure modes addressed in other products. Improvement opportunities identified as stresses and use profiles change.	Identified failure mechanisms addressed in all products. Advanced modeling techniques explored and adopted. Formal and effective lessons-learned process exists.	New technologies evaluated and adopted to improve reliability. Design rules updated based on field failure analysis.
Still learning more about reliability management. Becoming supportive and helpful.	Full participation. Understanding of absolutes of reliability management. Recognizing their personal role in continuing emphasis.	Consider reliability management an essential part of the company system.
Reliability manager reports to top management, with role in management of division.	Reliability manager is an officer of the company, reporting effective status and devising preventive action and involved with consumer affairs.	Reliability manager is on the board of directors. Prevention is the main concern. Reliability is a thought leader.
Warranty, corrective action materials, and engineering costs monitored.	Customer and lifecycle unreliability costs identified and tracked.	Lifecycle cost reduction done through product reliability improvements.
"Through commitment and improvement, we are identifying and resolving our problems."	"Failure prevention is a routine part of our operation."	"We know why we do not have problems with reliability."

Table 3.1. The reliability maturity matrix.

process, and management. These categories can be adjusted to the needs of individual companies. The columns represent the five stages of maturity. The reliability maturity matrix is presented in Table 3.1. (The matrix is available to download https://accendoreliability.com/go/pre. You can find the download on the *Reliability Maturity* page along with the ebook about the maturity matrix and how to apply it properly for your program.)

The concept of a maturity model is not new (Crosby 1980, Bollinger and McGowan 1991, Brombacher 1999). It provides a means to identify the current state and illuminate the possible improvements to a reliability program. The matrix serves as a guide to assist an organization in improving its program.

3.2 Application of the reliability maturity matrix

An organizational reliability gap assessment (see Chapter 6) is only of value when the resulting actions create a more effective (mature) reliability program. Moving to the right, or increasing maturity, on the matrix provides value to the organization. Some sources of value include reduced field failures, reduced cost of product development and testing, increased ability to hit market introduction deadlines, and increased market share.

Each organization's culture, history, capabilities, and priorities will influence any reliability improvement program. Local effective change management and the internal influence of thought leaders will also affect any improvement effort. Therefore, any effort to improve an organization's reliability maturity must account for the local culture and norms; consequently, each improvement program will be different. Yet, the basic tools, approaches, and processes related to reliability engineering do remain largely the same across organizations. The particular product and market may place unique constraints on specific tools, but the basics tend to remain consistent.

The reliability maturity matrix, detailed in this chapter, is referred to as the maturity model or matrix or as reliability maturity. The IEEE standard 1624 Standard for Organizational Reliability Capability (IEEE 2009), Crosby's *Quality Is Free* (1980), and other works (e.g., Schenkelberg 2016) related to reliability engineering provide further guidance. The intention of the matrix is to provide the recommended tasks to facilitate transitioning from one maturity level to the next across each measurement category.

In general, organizations tend to have consistent reliability maturity across categories. There may be some variation, yet commonly only one level higher or lower than the overall average maturity is seen. The maturity matrix consistency reflects the cultural elements and the overall organization's approach or policy toward reliability. The consistency also reflects the interconnectedness between categories.

Assessment is the tool used to identify an organization's maturity level and cultural aspects clearly. The recommendations generated by the assessment focus on reinforcing strengths and improving weaknesses. Also, the specific recommendations focus on moving the average maturity to the right or upward in maturity. Given the interconnected nature of the categories, it is often difficult to only improve one category to a higher maturity without affecting related categories.

3.3 Descriptions of stages

The reliability matrix has five stages of maturity. Generally, the higher stages are most cost-effective and efficient at achieving higher rates of consistent product reliability performance. These stages are uncertainty, awakening, enlightenment, wisdom, and certainty (see Figure 3.1).

The amount of effort and resources devoted to reliability activities changes per stage. Stage 1 (Uncertainty) is marked by very little to no effort related to reliability, whereas in stage 5 (Certainty) investments are made only in the necessary activities that provide the most value. In stage

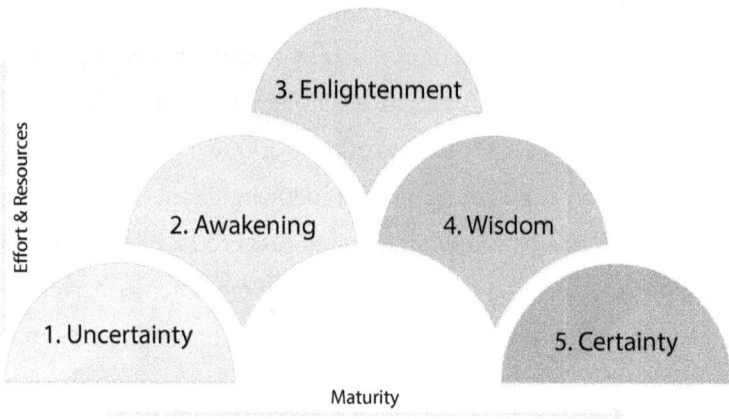

Figure 3.1. The five stages of reliability maturity.

3 (Enlightenment) a wide array of reliability methods are employed, often with results providing very little value.

Now let's discuss the traits (characteristics) that describe each stage one at a time:

Stage 1: Uncertainty. "We don't know why we have problems with reliability." Reliability is rarely discussed or considered during design and production. Product returns resulting from failure are considered a part of doing business. Field failures are rarely investigated, and often blame is assigned to customers. The few people who consider reliability improvements gain little support. Reliability testing is done ad hoc and often just to meet customer requirements or basic industry standards.

Stage 2: Awakening. "Is it absolutely necessary to always have problems with reliability?" Managers discuss reliability but do not support reliability activities with funding or training. Some elements of a reliability program are implemented, yet generally not in a coordinated fashion. Some tools such as FMEA, ALT, and HALT are experimented with, but most effort still focuses on standards-based testing and meeting customer requirements. Some analysis is done to estimate reliability or understand field failure rates, yet limited use is made of these data in making product decisions. There is,

48

however, an increasing emphasis on understanding failures and resolving them. Failure analysis is typically accomplished by component vendors, but results are of little value.

Stage 3: Enlightenment. "We are identifying and resolving our problems through commitment and reliability improvement." A robust reliability program exists and includes many tools and processes. Generally, significant effort is directed to resolving prototype and field reliability issues. Increasing reliance is placed on root cause analysis to determine appropriate solutions. Some tools are not used to their full potential because there is a lack of understanding of reliability methods and how the various tools apply. Some reliance is placed on establishing standard testing and procedures for all products. Only some use of these testing results is made for estimating product reliability to supplement predictions. Predictions are primarily made to address customer requests and not as feedback to design teams.

Stage 4: Wisdom. "Failure prevention is a routine part of our operation." Each product program or project has a well-developed reliability program that can be adjusted as the understanding of product reliability risks changes. Reliability tools and tasks are selected and implemented because they will provide needed information for decisions. Testing focuses on either discovering failure mechanisms or characterizing failure mechanisms. Testing often proceeds to failure, if possible. Advanced data analysis tools are regularly employed, and reports are distributed widely. There is increasing cooperation with key suppliers and vendors to incorporate the appropriate reliability tools upstream.

Stage 5: Certainty. "We know why we do not have problems with reliability." Product reliability is a strategic business activity across the organization. There is widespread understanding and acceptance of DFR and how it fits into the overall business. Product reliability is accurately predicted prior to product launch using a mix of appropriate techniques. New materials, processes, and vendors are carefully considered for their ability to meet internally established reliability requirements. The few failures that do occur are expected, and analysis is done to identify early signs of material or process changes. Customers and suppliers are regularly consulted on ways to improve reliability.

The stages of maturity may or may not proceed in a progression within an organization. It is not like a flower that begins as a seed and eventually matures and blooms. An organization may start at stage 2, skip stage 1, and never progress. Some organizations do advance from lower stages to stage 5. They also may regress to a lower stage over time. It is only with deliberate effort that an organization advances and maintains one of the more mature stages. Once stage 4 or 5 becomes established in an organization, the self-sustaining nature of those stages will take little effort to maintain.

3.4 Descriptions of matrix categories (rows)

The matrix rows are grouped into four categories of related subcategories or threads found in every product development and manufacturing organization. Having conducted numerous maturity assessments and mapping organizations to the matrix, the authors are keenly aware that using a specific reliability method (i.e., HALT or FMEA) is not an indicator of maturity. Instead, how and why each method is employed reflects the organization's maturity.

Let's examine just one method: HALT. An organization operating at any stage of maturity may be aware of HALT and may or may not conduct HALT studies. An organization operating at stage 1 may consider HALT a waste of time as it only creates failures, often only when the product is operated outside of normal operating conditions; thus, results are ignored, if even the testing is conducted. In stages 2 and 3, experimenting or always conducting HALT may take place, but benefits may be marginal because of poorly done studies, not conducting a thorough failure analysis, or not addressing issues identified. In stages 4 and 5, HALT is used only when it is the most suitable and valuable option to address the current situation, and action is always taken to resolve issues promptly and systemically.

Another pattern across the stages for a row is the sophistication of the employed methods. For example, less mature organizations use rudimentary data analysis methods, such as plotting averages or attribute data. More mature organizations may use regression analysis, design of experiments (DOE), and modeling approaches. The most mature organizations approach data analysis with the appropriate set of analysis tools for the specific task at hand, thus focusing on learning what needs to be learned from the data.

The matrix and expanded descriptions below the mentioned tools and methods illustrate the approach found for each stage. Of course, specific organizations may exhibit variations in how they employ specific or similar tools yet, in general, operate within a specific stage of maturity for a specific row.

3.4.1 Requirements

Whether the requirements and expectations are internally or externally created, each organization has its own set. Internally imposed requirements may define the desired characteristics of the product to create, product life-cycle milestones, or standing requests between groups within the organization. For example, customers and governments may impose functional performance and safety requirements. Training may be in response to requirements or part of maintaining (or improving) an organization's culture.

Requirements and planning. Designing and producing a product that meets customer expectations requires some level of understanding of customer expectations for functionality, use and environmental conditions, and durability. These requirements influence every facet of product design and production. The overall plan to achieve the reliability requirements establishes the sequence of reliability activities and decision points over the product life cycle.

Training and development. The technical skills and knowledge needed to design and produce a product span a wide range of reliability engineering activities. Individuals across the organization must understand the reliability-related goals, plans, tasks, and measures and their importance to effectively creating a reliable product.

3.4.2 Engineering

The term engineering here is inclusive of all individuals charged with identifying and solving problems. Design engineers include electrical, mechanical, software, and industrial engineers. Supply chain, sustaining, manufacturing, and maintenance divisions may also have engineering talent involved with reliability. The work of creating a reliable product is not the sole domain of the reliability engineer. In fact, many organizations do not have a single person identified as a reliability engineer, and some organizations still perform the appropriate reliability tasks and create highly reliable products.

Analysis. Assessing reliability risk with a product's design or field performance illuminates failure modes, mechanisms, and effects. The analysis provides information to create product lifetime estimates and predictions. Understanding, characterizing, comparing, and judging product reliability enables decisions across the product life cycle.

Testing. The intent of physically evaluating product prototypes and production units is to identify design and supply chain weaknesses, explore product limits and potential failure modes, and determine the effects of the expected range of use profiles and environments. Physical testing includes demonstrating that the product's durability (expected reliability) meets the requirements.

Supply chain management. Many products consist of a combination of purchased components and materials assembled into a functional item.

The overall reliability performance is significantly influenced by the reliability performance of the selected components and materials. Reliability is only one aspect of supplier selection, and the active involvement of reliability practitioners enables risk assessment, reliability requirements allocation, joint component reliability testing, and key vendor process control enhancements. Furthermore, monitoring supplier impact on reliability performance, process variation, change notices, and end-of-manufacture notices enables active management of any effects on product reliability.

3.4.3 Feedback process

There are many processes implemented to gain insights into what is working or what is not in terms of reliability. From simulating or testing prototypes to analyzing why something has failed, we often have some procedures in place to identify, track, and resolve failures. We also often use this information to make improvements, then set checks to ensure the "fix" actually resolves the problem as expected.

Failure data tracking and analysis. Each product failure highlights an area for product reliability improvement. Systematically recording, tracking, analyzing, and reporting failures from across the product life cycle and supply chain enable you to acquire comprehensive and timely information. The product design team needs to understand, prioritize, and design products to minimize product failure. The entire business requires timely and accurate failure data for decisions to be made concerning, for example, improvement projects, supplier selection, and warranty policies.

Verification and validation. This set of check steps in most organizations consists of verifying that the reliability objectives have been met and that planned reliability activities have occurred. A cross-check can support individual results with consistent results from other reliability activities. The process is often part of the overall program management process.

Improvement. During this process, one tries to identify and implement product changes that are designed to improve product reliability. The sources for improvement projects may come from reliability testing and analysis; product failures; customer requests; changes in the supply chain, use, or environmental conditions; or changes in technologies or materials. The implementation of corrective actions includes prioritization, validation of effectiveness, and prevention of recurrence of similar failure modes or mechanisms.

3.4.4 Management

The management team sets the tone for all aspects of an organization. The policies, practices, and priorities convey the management team's placement of reliability importance relative to the many priorities within the organization. How the management team acts is more important than the slogans or official statements—where is the attention and follow-up?, where are the resources being directed?, who is rewarded?, and what garners personal involvement?

Management sets the culture (the norms around behavior and decision-making) either inadvertently or deliberately. One approach is to establish a culture of "finding problems" (Yoshimura undated). However, it is the set of norms around how to identify, understand, avoid, and resolve problems that vary according to the maturity of the organization.

Understanding and attitude. This attribute is a reflection of the level of the management team's comprehension of reliability engineering's role within the organization. It is not limited to a person with the title of "reliability engineer" but rather encompasses all individuals who implement reliability methods and the level within an organization where reliability is a regular part of the agenda and decision-making.

Status. Within an organization, who are the leaders (independent of position)? What combination of voices tends to drive the company? Who is held in high esteem, rewarded, and promoted? The status of the reliability practitioner may range from nonexistent, to an obstacle, to a necessary part of doing business, to a valued team member, or to a thought leader. Do people want to implement reliability methods because they are viewed as important and career-enhancing? The status of those identified as reliability practitioners is one indicator of the value placed and recognized related to reliability engineering activities.

Cost of unreliability. The language of business is money. What does the organization track and value, and how is it expressed? The actual measures, their accuracy, and their relevance to decision-making express the importance of product reliability within an organization.

3.4.5 Prevailing sentiment

The prevailing sentiment is a short phrase that represents the mindset of those in an organization for each stage of maturity. Of course, within every organization there will be individuals with different sentiments, yet overall the culture of the organization tends to reflect a common sentiment.

3.4.6 Matrix summary

Identifying where an organization lands on the matrix is admittedly subjective. Nonetheless, by asking a selection of individuals familiar with an organization (typically having gone through multiple product development cycles or projects), it is usually found that they tend to agree with each other on the organization's stage of maturity for each row and overall.

An aspect that is not conveyed by the maturity matrix is the increase in the interdependence between rows as the level of maturity increases. Let's examine the use of HALT to illustrate. Lower maturity organizations may conduct HALT based on the idea that it might be a good thing to do by the folks conducting reliability testing. The preparation and results involve or affect other groups within the organization in a minor way, if at all. In more mature organizations, HALT may be initiated by a notice of a change in a supplier's process, include inputs from marketing on updated market use conditions and stresses, and involve cross-discipline support to assess and resolve any discovered weaknesses. As an organization advances in reliability maturity, the communication of reliability-related information increases across and within the organization.

The matrix is a powerful tool to help identify the current maturity of an organization. It may also assist in illustrating what a more mature condition may be like. Mapping maturity along each row helps to identify areas doing well and areas needing improvement.

3.5 Structure of the matrix stages: Reactive and proactive

Do you let events happen to you, or do events follow your designs and expectations? Are you a spectator or actor? Do you wonder about your product's future, or do you control your product's future? Are you reactive or proactive?

In practice, no organization is completely reactive or proactive. Within a typically reactive team, someone may identify a potential failure yet hesitate or be prevented from taking action to avoid future failures. When identifying and resolving a field failure, the same team may take steps to avoid similar problems in the future.

Predominantly proactive organizations may still discover an unanticipated failure mechanism and have to respond to address the problem reactively. The same team may also actively work to discover potential failure mechanisms and then take steps to address them in the current design.

Every reliability and maintenance program comprises a system. Every program has inputs, such as product testing results and field returns. Every reliability program has outputs, such as product design and production. A comprehensive reliability program includes product specifications for functionality and expected durability in the most basic terms. The program includes some form of design, verification, production, and field performance. Given this basic life-cycle description, it is possible for both reactive and proactive approaches to be employed depending on the specific situation within the program. The more mature teams are predominately proactive and are reactive as needed.

3.5.1 Every product will fail

Let's consider the notion that every product will eventually fail. Even the most robust product on Earth will fail when the Sun expires. Well before the collapse of the solar system, most products made today will have completely failed. The failures will range from deterioration of materials to overstress conditions (i.e., a lightning strike) or simply misuse. Some products will simply wear out; others will become obsolete and lose compatibility with other systems; others will simply no longer provide sufficient value.

Another important notion in product design is that there are a finite number of faults in the design. A push button has a limited number of actuation cycles before accumulated stress cracks the switch's dome spring. Material has a degradation mechanism (e.g., corrosion or polymer chain scission) that slowly deteriorates the material's strength. A bug in the software can disable the equipment temporarily. Furthermore, there are possible defects designed into the product that do not account for production variation, user demand, or environmental variations or do not anticipate user expectations. In every case, the design flaw will lead to failure sooner or later. Nonetheless, given only a finite number of failures, it is possible to find and remove most design errors.

3.5.2 Reactive approach

An approach to product reliability, and the most common method found in maturity stages 2 and 3, is to wait for product failures and then respond with analysis, adjustments, and refinements to improve product reliability. The naive method is to wait for customer failure reports before acting. The team's logic, if even considered, is the following: "We are good designers."

The customer will use the product in unforeseen environments and applications. If there are customer failures, we will consider improvements. For some products, with limited release and ample time to redesign the product, this may be perfectly feasible.

The design team could consider simple improvement by estimating the customer's use profile and environmental conditions. Armed with this information, the team then evaluates the impact of the conditions on the product's reliability through standardized testing. Setting testing conditions at or slightly above expected operating environments enables direct evaluation of the design to meet expected conditions. The faults found would be similar to the failure expected to occur in the customer's hands, and there may be time for a redesign before the product is shipped to customers. However, following this logic may lead to a broad spectrum of expensive and time-consuming testing.

Part of the logic of product testing includes the thought, "If we test in enough ways over the full range of use and environmental conditions, we should find and correct every design fault." There is often a heavy reliance on industry standards and common test methods for every product. Further improvements to product reliability can refine this reactive method and include using simulations, risk analysis, and early evaluation and testing of subsystems and components. The overall approach is often limited by the knowledge of actual use conditions, lack of test samples, and lack of time.

3.5.3 Proactive approach

Moving to a proactive approach, primarily in stages 4 and 5, can enable the reduction of product testing and increase product reliability. Although this may seem similar to the above approach, it involves a focus on failure mechanisms instead of test methods. Products fail because they do not have sufficient strength to withstand a single application of high stress (e.g., being dropped or a static discharge), or they accumulate damage (e.g., through wear, corrosion, or drift) with use or over time. By thinking through how a product could fail by considering the materials, design, assembly process, and these same aspects for vendor-supplied elements, the product team can construct a list of possible failure mechanisms.

In this approach, not all the failure mechanisms will be fully understood or characterized. The risk, in this case, is the decision to launch the product while not understanding the possibility or potential magnitude of product failure. The amount of risk itself is unknown. Therefore, the proactive team proceeds to characterize the design or material under the expected use conditions. The intent here is to reduce the uncertainty of the risk.

A second result of the proactive approach to risk assessment is the rank ordering of failure mechanisms by the expected rate of occurrence. One way to accomplish this ranking is to evaluate the stress versus strength relationships. Items with the greatest overlap of the two distributions (stress and strength) will have the highest potential for failure. The solutions may include increasing strength or reducing the variance of the strength.

A third result of the risk assessment is similar to the stress and strength evaluation and includes the impacts of time or usage on the change in the stress and strength distributions. Either curve may experience changes to the mean or variance over time. This may be due to degradation, wear, or increased expectation of durability by customers.

The proactive approach takes more thinking and understanding of how applied stresses create failures. Product designs, materials, and processes must be characterized, and their related failure mechanisms must be identified. This upfront effort offers rewards in terms of saved time and resources and avoiding redesigning and reworking.

In summary, in a reactive approach, one creates a design and then waits for field returns or standard product testing failures to prompt product improvements. In contrast, in a proactive approach, one anticipates failure mechanisms via expert knowledge, experience, experiments, or simulation and characterizes the response of the design and materials to expected stresses and then designs.

There are other aspects that differentiate between reactive and proactive reliability programs. For example, if management only discusses product reliability when a major customer complains about product failures, that is a reactive approach. If the management team regularly inquires and discusses the risk a particular design presents to reliability performance, that is a proactive approach.

Understanding the basic approach in use within an organization, in other words the reliability maturity, is one input to building a suitable reliability plan. It provides a means to identify strengths to amplify and weaknesses that may impede achieving stated objectives. Furthermore, understanding the current state of reliability maturity helps to communicate potential organizational or structural areas for improvement. If an organization moves from reactive to proactive approaches, or from lower to higher stages of maturity, the organization tends to anticipate and address reliability issues and enjoys the associated benefits.

3.6 Summary

Of importance is not which methods a team employs related to creating a product but rather why and how they use those methods to inform the decisions resulting in the product's reliability. How organizations address reliability varies from ignoring reliability completely and being surprised by the results (good or bad) to incorporating proactive reliability thinking throughout the organization and achieving expected results with few, if any, surprises.

If your organization would benefit by operating at a higher maturity level, then the next chapter will be a great value. It will introduce the necessary steps to both create a highly reliable product and improve the culture and the corresponding reliability maturity level.

Chapter 4

PRIMARY STEPS TO ACHIEVING HIGH RELIABILITY

"Would you tell me, please, which way I ought to go from here?"
"That depends a good deal on where you want to get to."
"I don't much care where...."
"Then it doesn't matter which way you go."
Lewis Carroll, Alice in Wonderland

In this chapter. We start with a brief discussion of the connection between customer expectations and a suitable set of reliability objectives. Next, we introduce the thinking behind the six steps necessary to achieve high reliability. We then finish the chapter with an examination of each step and when to use this process.

4.1 The goal of achieving high reliability

Every product needs to have high reliability ... right? Not necessarily. Depending on customer expectations and product cost trade-offs, some products are designed to have higher reliability than others.

In the bicycle Pro Series, the bicycle is designed for road bike racing teams and serious amateurs. Teams generally use new bicycles each season. Serious amateurs may use the equipment for two or three years. Reliability and safety during the bicycle's shorter life are very important. A broken bicycle will take you out of the race, not to mention the possibility of serious injury.

In the bicycle Intro Series, the bicycle is designed for those new to cycling or who need an economical bicycle solution. The intended useful life is 5 to 10 years. Riders of the Intro Series understand that they have made cost trade-offs and do not expect the highest level of reliability. Safety is important, but customers may be willing to put up with smaller discomforts or inconveniences in exchange for the lower price.

When setting reliability requirements for new products, project teams consider many factors. In the bicycle example, what kind of bicycle is planned? Is it for first-time riders or experienced racing teams? What are the materials and usages? What are the expectations for useful life? What are the design focus areas?

Customer expectations drive reliability objectives.

4.2 Rationale for each step

We'll begin this section with an analogy, staying with the bicycle example.

Let's say you are a reliability engineer on the Pro Series bicycle team and designing a brand-new racing bicycle. The team wants to design and build a reliable bicycle, and you are charged with supporting that reliability. Where do you begin? Would you begin by selecting materials? Probably not, as you haven't yet decided on the requirements for the new bicycle. Would you begin with a physics-of-failure model? Probably not, as you may not have the skills or the time to do such an analysis, and we don't even know whether this type of analysis is needed. So, where would you begin?

In *The 7 Habits of Highly Effective People*, American educator Steven Covey (2020) advises, "Begin with the end in mind." What kind of bicycle do we want? What are customer expectations? What are the specifications or requirements? Does the new bicycle need to be more reliable than that of the competition? The first step in designing a reliable bicycle is defining the vision and the requirements for reliability. We call the first step "Develop a reliability strategic vision."

Once we have outlined the reliability vision for the new bicycle, including requirements and maybe even sketches or drawings, what's next? Do we start right in designing the bicycle? Probably not. We need to raise questions and assess resources and capabilities. Does the Pro Series team have experience with designing reliable bicycles? What are the biggest challenges or field problems identified from past racing bicycles? What methods, such as FMEA or HALT, does the team have experience with, and what have been the results from past analyses? What are the gaps that need to be addressed to achieve the objectives? The second step in designing a reliable bicycle is to do an assessment of the resources and capabilities. We call the second step "Perform a reliability gap assessment."

At this point, we have our vision (including requirements) and have realistically assessed the resources, skills, and capabilities that will be needed to achieve our vision, including any deficits we need to address. What's next? Reliability doesn't occur in a vacuum. Achieving reliability on projects or programs requires supporting key decisions. The key project or program decisions advance product development, and an important step in achieving high reliability is supporting the most important decisions. We call this third step "Identify reliability-related decisions."

By this time, we should be able to identify the tools and methods that will be needed to get the job done. What specific reliability methods will close the gaps and achieve reliability objectives? We call the fourth step "Select the right reliability methods."

We are ready to establish a plan of action to design, test, and build a reliable new bicycle. Do Task A, Task B, and Task C. Depending on the complexity, we might identify subtasks to define the needed action steps clearly. Each task includes what, who, when, and how. The detailed plan of action is the fifth step. We call this step "Create an effective reliability plan."

Once the plan of action is established, we can begin executing the action steps, including hiring the appropriate support staff, conducting needed training, performing the various reliability methods, and ensuring that each step is properly executed. The sixth step entails executing the plan of action to design, test, and build a reliable bicycle. We call this step "Execute reliability plan tasks."

All of these steps to achieve high reliability are done in parallel to the product development process. In fact, the best companies integrate the steps for achieving high reliability into their product development process work instructions. Reliability is not a "program of the month" and is not achieved by placing banners on the wall. It entails an ongoing process.

4.3 Description of the six steps to achieving high reliability

In this section, we'll briefly describe the six steps to achieving high reliability (see Figure 4.1). Each of these steps has its own chapter in the book. Here we provide an overview.

Step 1: Develop a Reliability Strategic Vision. It is important to begin any new activity with a clear understanding of the goals and objectives. Reliability strategic vision is the envisioned future from a reliability viewpoint. It includes the necessary infrastructure, resources, skills, methods, capability, and reliability of products and processes. It is developed with

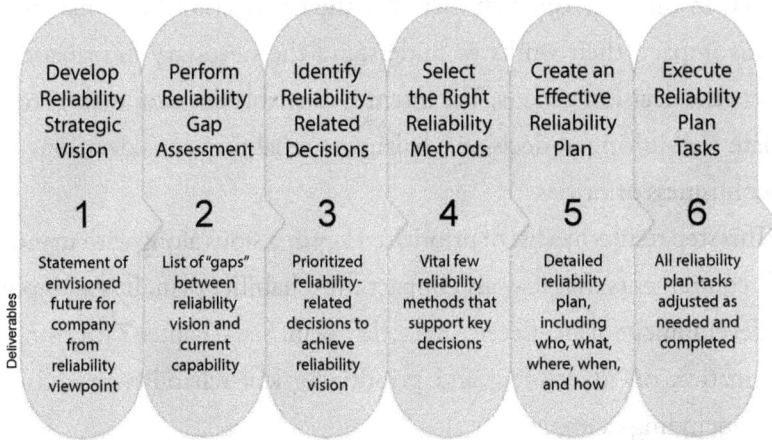

Develop Reliability Strategic Vision	Perform Reliability Gap Assessment	Identify Reliability-Related Decisions	Select the Right Reliability Methods	Create an Effective Reliability Plan	Execute Reliability Plan Tasks
1	2	3	4	5	6
Statement of envisioned future for company from reliability viewpoint	List of "gaps" between reliability vision and current capability	Prioritized reliability-related decisions to achieve reliability vision	Vital few reliability methods that support key decisions	Detailed reliability plan, including who, what, where, when, and how	All reliability plan tasks adjusted as needed and completed

Deliverables

Figure 4.1. High-level primary steps to achieving high reliability.

full management support and understood by all employees. See Chapter 5 for a full description of reliability strategic vision, including how it is developed and examples.

Step 2: Perform a Reliability Gap Assessment. Reliability gap assessment begins with understanding the reliability vision. This assessment is done via observations and direct communication with key stakeholders in the company. Shortcomings (gaps) are analyzed to achieve the vision, the capabilities of current products and processes are identified, and the effectiveness of current reliability methods are assessed. Gaps include both organizational capability and the application of reliability methods.

From a business and technical standpoint, the assessment document outlines the shortcomings or gaps between reliability vision and current reliability capability. It is an input to the development of an effective reliability plan. See Chapter 6 for complete information on how to perform a reliability gap assessment, including examples.

Step 3: Identify Reliability-related Decisions. Reliability occurs at the point of decision. Although reliability may influence the many decisions that establish the reliability performance, not all decisions have the same impact. In this step, one first identifies the key decisions that may have a serious impact, then works to understand the necessary information to inform the decision makers. The intent is to enable decision makers to formulate well-informed decisions, balancing reliability considerations with other business priorities.

This step results in a list of prioritized key decisions along with mandates and requirements that may also impact the reliability plan. It is an input to the development of an effective reliability plan. See Chapter 7 for complete information on identifying and prioritizing key reliability-related decisions, including examples.

Step 4: Select the Right Reliability Methods. The list of prioritized reliability-related decisions, along with the reliability strategic vision and the reliability gap assessment, guides the selection of the vital few reliability methods that are needed to inform decision makers, close the gaps, and achieve the reliability objectives. It is important to be specific about the method description. Don't just say "FMEA," say "Perform a design FMEA on subsystem X." Don't just say "DOE," say "DOE on the decal application process."

When selecting the appropriate reliability methods that will become the core of the reliability plan, it may be useful to consider reliability method categories as thought-starters. See Chapter 8 for complete information on selecting the right reliability methods. Within Chapter 8, see Section 8.3.2 for a description of the categories for reliability methods.

Step 5: Create an Effective Reliability Plan. At this point, we have identified the vital few tools that are needed to achieve the reliability objectives. We are ready to develop the specific tasks that close the gaps from the reliability gap assessment and achieve the vision for reliability. The needed tasks pull together all necessary resources, address high-risk areas, strengthen organizational shortfalls, and utilize the selected tools. We transition from a list of reliability tools to specific tasks (what, who, when, where, and how). The plan is organized and approved by management.

See Chapter 9 for complete information on how to create an effective reliability plan, including examples.

Step 6: Execute Reliability Plan Tasks. If you have completed the first five steps properly, the reliability plan will encompass a set of tasks that are necessary for project success and have been bought into by management. The tasks will be described in sufficient detail to accomplish the objectives and written in an executable manner. At this point, the focus is on execution, and the key to execution is to leverage management support to ensure that each task is completed, all gaps are closed, and reliability objectives are achieved.

See Chapter 10 for complete information on how to execute reliability plan tasks, including examples.

Project teams have proven the above process steps over and over to be the most effective way to achieve high reliability. In the next section, we'll discuss when the six steps are needed and when they can be shortened.

4.4 When to use the six-step process

Applying the six-step process must include product development considerations and a market-specific set of constraints. Constraints may be technical, financial, geopolitical, etc. Therefore, each plan may include different types of tools and tasks to achieve specific objectives.

Note that not every reliability project needs to go through these six steps. We'll outline two examples to illustrate when the steps are needed and when they can be shortened.

The first example is from the Intro Series. Recall the Intro Series is a high-volume, low-cost bicycle for new or economy-minded bicyclers. Let's say the project team is making a minor revision to the handlebar geometry to facilitate easier assembly and reduce cost. We would characterize this as a minor project. The project team would not necessarily need to go through all six steps of the six-step process. In this case, the team might go straight to step 3 and select the tools needed to ensure that the new geometry of the handlebar meets reliability objectives, such as change point analysis, an updated FMEA, and validation testing, followed by execution of the selected tools. A full reliability plan would not be needed

Consider the scenario of a new Pro Series bicycle model as the second example. Recall that the Pro Series is a low-volume, high-cost road bike designed for racing teams. Let's say the Pro Series team is developing a new model incorporating the latest technology and materials for lightweight frames and improved braking and suspension. We would characterize this as a major project and we should go through all six of the six steps to achieve high reliability. A comprehensive reliability plan is needed here.

Potential reliability risk is present in both cases. In the first example, the reliability risk can be addressed with a shortened version of the six steps, focusing on the potential failure mechanisms related to the handlebar geometry change. In the second example, the reliability risk is broader across more of the bicycle and encompasses more potential failure mechanisms that need to be addressed with a comprehensive reliability plan.

When facing a new project with the potential for significant risks, such as new technology or new applications, that is the time to use the six steps to achieve high reliability. Using the appropriate steps in the six-step process will help achieve robust designs that meet reliability objectives.

4.5 Summary

This introduction to the six-step process provides a brief overview. From understanding the desired objective to execution, the process relies on having the right information to craft the plan. The following six chapters will explore each step, in detail, starting with step 1 on developing a reliability strategic vision.

Chapter 5

DEVELOPING A RELIABILITY STRATEGIC VISION

The greatest danger for most of us is not that our aim is too high
and we miss it, but that it is too low and we reach it.
Michelangelo

In this chapter. The first of the six steps is gaining an understanding of where senior management intends to take the organization related to reliability performance. With that information in hand, it is then possible to craft project specific reliability objectives or goals.

5.1 What is the reliability strategic vision?

From a reliability viewpoint, where does your company want to be in two years, three years, or five years? Consider the following:

- Does your company want to become an industry leader in quality and reliability?
- Is the goal to achieve fewer than one defect per thousand systems in the first year after launch?
- Do you want to reduce warranty expenses by at least 50%?
- Should we expect fewer than 1% returns over the product's life.
- Does your company want to develop in-house reliability expertise to support all future projects?

71

A reliability strategic vision is the envisioned future for the company from a reliability viewpoint. It follows one of the basic premises of management: Identify where you want to go, then develop a systematic and structured plan to get there. It is not a slogan emblazoned on a banner or a wall placard. It often requires a genuine culture change.

As mentioned before, Covey (2020) advises, "Begin with the end in mind." The reliability strategic vision outlines the overall vision for reliability for the program or project. It should be developed with full management support, understood by all employees, and integrated into work activities.

To understand the concept of reliability strategic vision, it is helpful to begin with some basic definitions.

One way to define the word "strategy" is "the determination of the basic long-term goals and objectives of an enterprise and the adoption of courses of action and the allocation of resources necessary for carrying out these goals."

The phrase "strategic vision" may be defined as "ideas for the direction and activities of business development, generally included in a document or statement so that all company managers can share the same vision for the company and make decisions according to the shared principles and company mission."

Collins and Porras (1996) explain that

> Truly great companies understand the difference between what should never change and what should be open for change, between what is genuinely sacred and what is not. This rare ability to manage continuity and change—requiring a consciously practiced discipline—is closely linked to the ability to develop a vision. Vision provides guidance about what core to preserve and what future to stimulate progress toward. A well-conceived vision consists of two major components: core ideology and envisioned future.

The scope of the reliability strategic vision includes personnel, training, product, process, policy, methods, and assets. It is more than a reliability metric, although numerical reliability goals can be part of the strategic vision for reliability. Imagine what the company needs to be, the infrastructure, skills, methods, capability, quality, and reliability of the products and processes, and when that transformation will be completed. That is the essence of reliability strategic vision.

For a specific project, the scope embraces the reliability vision for that specific project. The scope can also be an entire program, such as a series of product lines.

5.2 Where does the reliability strategic vision fit into the six-step process?

The reliability strategic vision is the first of the six steps taken to achieve high reliability. It is an important input to set the specific reliability requirements covered below and to perform a reliability gap assessment, which is the subject of Chapter 7.

Primary Steps to Achieve High Reliability
(High Level)

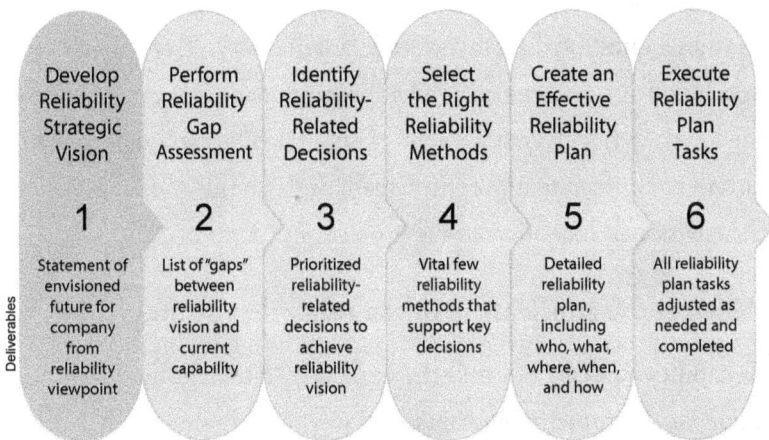

Develop Reliability Strategic Vision	Perform Reliability Gap Assessment	Identify Reliability-Related Decisions	Select the Right Reliability Methods	Create an Effective Reliability Plan	Execute Reliability Plan Tasks
1	**2**	**3**	**4**	**5**	**6**
Statement of envisioned future for company from reliability viewpoint	List of "gaps" between reliability vision and current capability	Prioritized reliability-related decisions to achieve reliability vision	Vital few reliability methods that support key decisions	Detailed reliability plan, including who, what, where, when, and how	All reliability plan tasks adjusted as needed and completed

Deliverables

5.3 What questions need to be answered to develop a reliability strategic vision?

Management should work with program or project stakeholders to develop the vision for reliability. Answering questions similar to the following can be helpful:

- What are customer expectations for reliability?
- What are the operating conditions that represent the most severe usage anticipated?
- What are the operating conditions that represent the typical usage anticipated?
- What are customer expectations for how long the product should last in service?
- How should the product or brand be positioned relative to the competition?
- What are the warranty cost objectives?
- What reliability infrastructure, including staffing, training, skills, and methods capability, is envisioned?
- What are the goals for supplier reliability?

Other questions that help you verbalize the envisioned reliability or understand the desired position of your product in the marketplace can also be beneficial.

As these questions are answered, a reliability strategic vision will emerge. It will also become clear what program decisions can be supported (influenced) with a clear statement of the reliability strategic vision. It is important to confirm that the entire team understands and feels ownership for the envisioned reliability of the project or program.

5.4 What is the application of the reliability strategic vision?

The reliability strategic vision is the company statement used to guide the direction of the project or program. For a specific project, it is the reliability vision for that specific project. For a program, such as a series of product lines, it is a reliability vision for the entire program. It includes the corporate or division reliability goals. Likewise, the project reliability goal is informed by the vision and corporate reliability objectives even for a specific product.

People need to know where they are headed when accomplishing any big objective. For example, if the goal is to climb a mountain, the climbing team needs to figure out which mountain range they will be climbing, which peak they will climb, how long it will take, and what envisioned climbing skills are needed.

The next step in the six-step process is to assess where the project or program stands currently relative to reaching the goal, the reliability gap assessment. A well-stated reliability strategic vision will be used as input to the next steps in the overall process to achieving high reliability.

Projects can have many other goals beyond reliability. Budget, timing, marketing, performance, and other project objectives all are important and can be connected in various ways to reliability. For example, product revenue and expenses affect profitability, and achieving high product reliability can increase sales and reduce field service costs. Another example is in the area of marketing: High reliability can be an excellent marketing advantage. A company known for high reliability has a distinct marketing advantage over the competition, which translates to the bottom line. Conversely, no amount of aggressive marketing or public relations can save a product or service that is not reliable.

Table 5.1 lists the findings of Gnanapragasam et al. (2018) from a survey conducted on consumer purchasing factors across 18 product categories. The survey of United Kingdom adults used a defined scale with options ranging from "not at all important" to" extremely important." The study focused on the relative importance of six different purchasing factors: appearance, brand, guarantee, longevity, price, and reliability. A summary from the paper concludes that "most consumers consistently emphasize the importance of longevity and reliability when purchasing new products."

Category	Appearance	Brand	Guarantee	Longevity	Price	Reliability
BICYCLES	V	M	V	V	V	E
CARS	V	V	V	E	E	E
CLOTHING	E	M	M	V	V	—
ELECTRONICS GOODS	M	V	V	E	V	E
FLOOR COVERINGS	E	M	V	E	V	—
FOOTWEAR	V	M	M	V	E	—
FURNITURE	E	M	V	E	V	—
HOUSEHOLD TEXTILES	E	M	M	V	V	—
JEWELRY, CLOCKS, AND WATCHES	E	M	V	V	V	E
KITCHENWARE	E	M	M	V	V	—
LARGE KITCHEN APPLIANCES	V	M	V	E	V	E
MUSICAL INSTRUMENTS	V	M	V	V	V	—
POWER TOOLS FOR HOME AND GARDEN	M	M	V	E	V	E
SMALL HOUSEHOLD APPLIANCES	M	M	V	V	V	E
SMALL TOOLS AND FITTINGS	M	M	M	V	V	—
SPACE HEATING AND COOLING PRODUCTS	M	M	V	E	V	E
SPORTS EQUIPMENT	V	M	M	V	V	—
TOYS AND GAMES	M	M	M	V	V	—

Legend: M   Moderate   V   Very   E   Extremely

Table 5.1. Median importance of purchasing factors. (Note that reliability was only assessed for products with complex electrical, electronic, or mechanical parts.)

5.5 Developing and using reliability requirements

In the reliability engineering and management field, there is no more important subject than how to develop and implement robust reliability requirements. This section will outline some of the key elements that need to be addressed in crafting such reliability requirements.

There are three sources of input when crafting reliability goals: customer expectations, business objectives, and technical feasibility. These all add credibility to the goal and provide a touchstone for every decision through the product development process.

Customer expectations. What is an acceptable chance of failure over what duration for the customer? What is the environment (where used) and how is the product used (i.e., the use profile)? Customers may state their expectations in the form of requirements, or they may have unstated expectations based on personal experience with similar products or based on warranty terms.

Business objectives. Customer satisfaction, market share, and profit intertwine with the reliability performance of a product. Businesses incur expenses when a product fails. These expenses include warranty, engineering, and brand reputation, and these expenses deduct from the realized profit.

Technical feasibility. Product reliability is just one of many factors that influence the final design of a product. Other factors include material capability, the laws of physics or chemistry, patent ownership, etc. Although it is technically impossible to create a product that will never fail, it is possible to create one that minimizes the chance of failure over the useful life.

5.5.1 What is reliability?

For discussion purposes, we'll use the definition of reliability from *Practical Reliability Engineering* (O'Connor and Kleyner 2012, p. 1):

"The probability that an item will perform a required function without failure under stated conditions for a stated period of time."

Let's consider these elements, one at a time:

1. The "probability" usually refers to the likelihood of successful operation.
2. The "required function" comprises the reliability requirements that should be associated with specific product functions.
3. The "understated conditions" are the reliability requirements that should include the conditions of use, both environmental stresses and customer usage patterns.
4. The "stated period of time" denotes the reliability requirements needed to specify over what duration the reliability is being measured or required.

Any reliability "requirement" that does not clearly include each of these four elements is not a well-written or fully stated reliability requirement.

5.5.2 Wrong way to specify reliability requirements

Here are a few examples of the wrong way to specify reliability requirements.

"The widget must be 99% reliable at ten years." What's missing? Required functions and stated conditions.

"The widget must achieve an MTBF of 100 hours." MTBF is not a probability; it is the inverse of an average failure rate. What's missing? Functions, duration, and stated conditions. Because of its many shortcomings, we do not recommend using MTBF or similar "mean time between" statements. See NOMTBF.com for more information concerning the perils involved when using MTBF.

"The widget must be 95% reliable when operated according to functions and conditions of use from document ABC." What's missing? The time element.

5.5.3 Right way to specify reliability requirements

Here is an example of the right way to specify reliability requirements:

> The Enthusiast road bicycle used on paved roads worldwide must be at least 99% reliable at 10 years of life while being operated according to the functions and conditions of use from document XYZ.

Each of the four elements of reliability requirements must be well thought out and specified.

The probability element should be consistent with the "reliability strategic vision" of your company. If the Enthusiast braking subsystem needs to be more reliable than that of the competition, consistent with warranty reduction goals of 30%, and achievable within the product development cost requirements, all of those objectives must be achieved with the statement of probability. Setting the probability statement too high could jeopardize product development cost objectives and require extensive verification testing; setting it too low would cause the subsystem to not exceed the competition's reliability or fail to meet warranty reduction goals.

Here is another example of reliability requirements for an Enthusiast Series braking subsystem:

When considering the function element, a brief story may illustrate. In a bicycle test lab, braking subsystems were tested years ago. The subsystem had to be 95% reliable. At the time, only hard failures were counted against this requirement. In other words, failure meant the brakes would not provide sufficient stopping power. A technician asked how much braking force is needed to stop the bicycle. The Enthusiast bicycle technical specifications stated that the brakes must provide at least 100 foot-pounds of stopping power within 3 seconds at 20 degrees Fahrenheit when the bicycle is operated at full speed. Merely stopping the bicycle is not sufficient by itself. The bicycle must stop in the required time and distance under the anticipated operating conditions. If the bicycle does not stop under these specifications, that is a failure. The intended function of the product must be part of the reliability requirement.

The conditions of use should be consistent with the intended environmental and customer usage requirements, as documented in technical specifications. What are the temperature, humidity, and dust profiles? How many cycles up or down are stipulated? What about vibration loads? These conditions of use are normally part of technical specifications. In the following car-door window example, one of the conditions of use would be operating at minus 40 degrees Fahrenheit.

The time element is also an essential part of every reliability requirement. Consider a car-door window, the target life for which is 10 years or 150,000 miles, whichever came first. The reliability requirement would be 95% reliability over 10 years (or 150,000 miles) while being operated according to the functions and conditions of usage outlined in document ABC.

Reliability requirements are more than exhortations proudly plastered on large banners. They are very specific statements that meet each of the four elements outlined above.

5.5.4 Example of reliability requirements for the Enthusiast Series bicycle

Using the information from the Enthusiast reliability strategic vision above, the following four elements of reliability requirements can be identified for the bicycle system:

1. Probability element: system reliability target of 95%.
2. Function element: Primary functions for the Enthusiast Series based on technical specifications document #123.

3. Conditions of use element: Specific operating conditions for the Enthusiast Series based on conditions-of-use document #456, based on the most severe conditions of the nine product lines.

4. Time element: The target life duration of seven years.

In summary, the system reliability requirements for the Enthusiasts Series bicycle is 95% reliability at seven years, based on the primary functions in document #123 and the conditions of use in document #456.

Reliability requirements for the Enthusiast Series subsystems and components can be obtained using reliability allocation methods, such as system reliability modeling.

5.6 Examples of reliability strategic vision

To illustrate the concept of reliability strategic vision, we'll use examples from each of the bicycle line scenarios from Chapter 1. Each of these examples describes the envisioned future for the Pro Series, Enthusiast Series, and Intro Series, respectively, from a reliability standpoint. This becomes the first step in the "Primary Steps to Achieving High Reliability."

5.6.1 Pro Series

Based on the interviews with the division management and the Pro Series project team, the following is a summary of information that becomes the reliability strategic vision for the Pro Series.

Pro Series customers participate in road bike racing teams and are serious amateurs. Weight, responsiveness, and overall performance are critical factors. Long-term reliability considerations are minimized, as teams generally use new bicycles each season. However, reliability and safety during the bicycle's shorter life are very important. Pro Series customers take the design, performance, and reliability seriously over the useful life of one season. Very harsh use, high power, all environmental conditions, daily pressure-washing, and tune-ups stress every bicycle component.

The target life duration for most components will be three years, with a system reliability target of 98% at one year, under the anticipated harsh operating conditions. The bicycle structure must be highly reliable with no failures in the first year of use.

The design team attempts to reduce weight, increase performance, and enable the bicycle to perform without failure (structural failure) and regularly considers accumulated-damage-type failure mechanisms to balance the chance of failure and performance.

Because of the shorter life cycle for the Pro Series line of bicycles and the need for high reliability in harsh conditions, the reliability strategic vision for the Pro Series includes a single highly experienced reliability engineer to coordinate the necessary design and assembly activities.

5.6.2 Enthusiast Series

Based on interviews with division management and the Enthusiast Series project team, the following is a summary of information that becomes the reliability strategic vision for the Enthusiast Series.

Enthusiast Series customers typically ride for fun, recreation, and fitness. They require moderate performance and reasonably good reliability, under usual operating conditions. The reliability vision of the Enthusiast Series team reflects these standards, with a system reliability target of 95% at seven years, under the prescribed operating conditions.

Although there will be up to nine product lines available during any given year, each having a wide range of operating conditions, the conditions of use profile used for the design and testing of every product line will represent the most severe conditions. Materials will need to be selected based on a balance of cost, weight, and reliability.

All bicycle systems will undergo normal testing and analysis, with a focus on safety and reliability. Suppliers will need to be selected based on a balance of cost and reliability. The target life duration will be seven years. Enthusiast Series customers must maintain their bicycles based on published preventive maintenance procedures.

Some members of the Enthusiast Series team will be trained on reliability methods, with monthly management reliability status reviews. The vision is for reliability training to be conducted by a reputable outside firm.

5.6.3 Intro Series

Based on interviews with the division management and the Intro Series project team, the following is a summary of information that becomes the strategic vision for the Intro Series.

Most users of the Intro Series are new to cycling or need an economical bicycle solution. To achieve this price point, the target life for the bicycle is five years, with a warranty set at two years. With minimum attention, shelter, and maintenance expected by users, selected parts should not need maintenance during the target life period.

The Intro Series will use mostly proven parts and involve fewer innovations and less new technology. Reliability testing will be folded into the larger design validation department, and there will be no need for dedicated reliability resources. Under the prescribed operating conditions, the system reliability target is 90% at five years.

5.7 Summary

A strategic reliability vision and a reliability goal are both descriptions of objectives. Having a clear understanding of the desired results benefits all in the organization.

In the next chapter, we will explore the reliability gap assessment step to understand the current strengths, weaknesses, and capabilities of the project or process.

Chapter 6

PERFORMING A RELIABILITY GAP ASSESSMENT

The wise man bridges the gap by laying out the path by means of which
he can get from where he is to where he wants to go.
J. P. Morgan

In this chapter. The next step in the process is to understand the organization's current capabilities and consider what else is necessary to achieve the reliability goals and vision defined in the first step. We introduce the idea of conducting a gap assessment including when and how to perform one, how to prepare and conduct the assessment, and how to analyze the results.

6.1 What is a reliability gap assessment?

Many companies write a reliability plan and attempt to implement reliability tasks without first understanding what drives reliability task selection. A reliability gap assessment aims to identify the shortcomings in achieving reliability objectives so that a reliability plan can be properly developed.

By definition, a reliability gap assessment is a comprehensive analysis of the specific gaps between a company's vision for reliability and the current reliability capability. It begins with developing or understanding the company's reliability vision, understanding the current reliability maturity, and then analyzing the specific shortcomings to achieving the vision. The "gaps" are those issues or shortcomings that, if closed or resolved, would move the company in the direction of achieving its reliability objectives. The assessment also reveals the necessary information to identify the reliability maturity of the company. The gaps and maturity include both organizational capability and application of reliability methods.

Although we use the term "reliability gap assessment," it is important to assess both strengths and weaknesses. Strengths are the areas where the current reliability capability is adequate to meet the vision for reliability. They should be reinforced and can be leveraged to achieve the overall reliability objectives.

The reliability culture can include reactive versus proactive tasks and inefficient versus efficient approaches. The culture, reflected by a stage of reliability maturity, provides a means to communicate the current state of the organization's reliability maturity. In general, one of the objectives is to move the organization's program toward an efficient, proactive culture.

Because the objective of a reliability plan is to focus on creating the vital information that is most effective and applicable to achieving safe and highly reliable products and processes, this approach specifically precludes enumerating a long list of tasks that may exceed resources and capabilities. The gap assessment helps to identify the short list of necessary information to be created by reliability tasks that will most effectively achieve objectives.

The reliability gap assessment can be done generically across all company projects and then modified to fit specific projects of concern. It can also be more narrowly applied to cover the scope of a given project. It is a key input to the reliability plan.

It should be noted that an assessment is not an audit. We are not assessing how well a reliability program meets a standard. Rather, we are assessing the capability of the current project and program team to meet the reliability goal or vision.

We have chosen to use the phrase "reliability gap assessment" rather than "reliability assessment." Reliability assessment often refers to the reliability metric and assesses the ability of the product to meet reliability requirements. Reliability gap assessment, in contrast, assesses the capability of the organization and its methods to achieve the vision for reliability, summarizing the barriers to achieving the vision.

6.1.1 When is it performed?

For a reliability program that involves a series of projects, the reliability gap assessment is done prior to devising the reliability plan, as it serves as one of the inputs. For a specific project, the reliability gap assessment should be done early in the product development process, such as when the product concept is first identified. As conditions within an organization change, the gap assessment may change as well. Therefore, periodically conducting a reassessment of what is or is no longer impeding achieving the reliability vision is warranted.

6.1.2 Where is it performed?

A reliability gap assessment is typically done onsite by observing product development and manufacturing operations, interviewing selected staff, and analyzing appropriate information. When deciding the scope of the assessment, the culture of the team members who work on the projects and divisions needs to be considered. Teams from different groups can have different cultures, and this affects their priorities and decision-making process concerning reliability.

Recall the bicycle Enthusiast Series. The Enthusiast division was purchased very recently and did not have an active reliability program. Compare that to the Professional Series, in which the lead designer is well known in the pro cycling world for the creation of highly reliable solutions, and the project team has reached a high level of reliability maturity.

You can group projects or organizations with similar cultures together when doing an assessment but one must assess projects or organizations with very different cultures separately.

6.2 Where does reliability gap assessment fit into the six-step process?

Performing a reliability gap assessment is the second of the six steps taken to achieve high reliability. It is an important input in identifying reliability-related decisions (Chapter 7) and selecting the right reliability tools (Chapter 8).

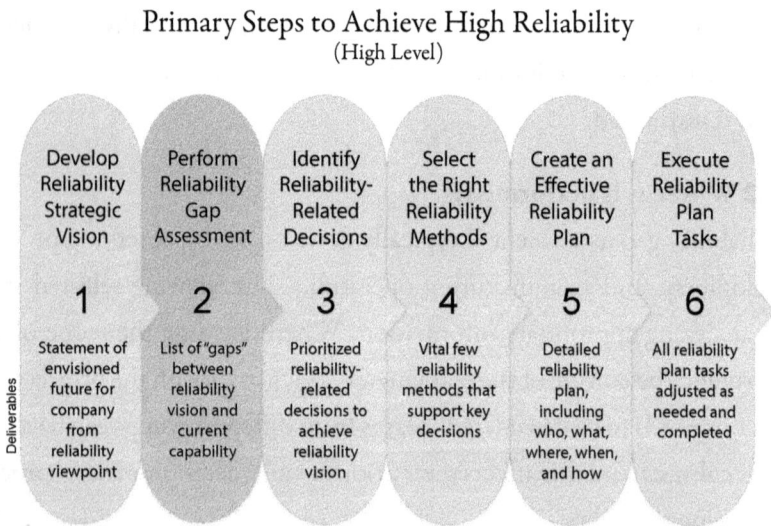

Primary Steps to Achieve High Reliability
(High Level)

Develop Reliability Strategic Vision	Perform Reliability Gap Assessment	Identify Reliability-Related Decisions	Select the Right Reliability Methods	Create an Effective Reliability Plan	Execute Reliability Plan Tasks
1	2	3	4	5	6
Statement of envisioned future for company from reliability viewpoint	List of "gaps" between reliability vision and current capability	Prioritized reliability-related decisions to achieve reliability vision	Vital few reliability methods that support key decisions	Detailed reliability plan, including who, what, where, when, and how	All reliability plan tasks adjusted as needed and completed

Deliverables

6.3 How is a reliability gap assessment performed?

The following outlines one way to conduct a reliability gap assessment. Any assessment will have to be specific to your local circumstances. Very few people within an organization take the time to fully understand the entire spectrum of current reliability-related activities, information, and decisions. When done well, the assessment provides an overall view of a complex process that involves many roles within an organization. Moreover, the assessment analysis provides a clear set of obstacles that impede creating a highly reliable product.

The reliability gap assessment has two distinct outputs: 1. It presents a clear understanding of the organization's reliability maturity graphically depicted using the reliability maturity matrix and 2. it provides a list of gaps that may impede the ability of the organization to achieve its reliability objectives.

6.3.1 Developing the list of reliability gap assessment questions

As covered above, when performing a reliability gap assessment, we are assessing the capability of the current project and program team to meet the reliability goal or vision and summarizing the most important gaps to the vision. This assessment is typically done onsite by observing product development and manufacturing operations, interviewing selected staff, and analyzing appropriate information.

When interviewing staff, it is helpful to prepare a list of questions for selected staff to be sure that nothing is missed. Asking the right questions can make the difference between a thorough and a cursory understanding of the issues and their causes. See Section 11.3 "Questioning" in Chapter 11 for guidance on how to construct meaningful and constructive questions. See Appendix A, online at accendoreliability.com/gap-assessment-questions, for a listing of potential questions that can be used as thought-starters to ensure coverage of the scope of the reliability gap assessment.

6.3.2 Example questions that may be useful as part of a reliability gap assessment

The example questions below are for the fictitious Enthusiast project. They are examples of how to engage and frame the discussion. Actual questions that are part of a reliability gap assessment must apply to the unique circumstances of the industry, company, and team culture, based on products, processes, marketing, cost, and other considerations. It is important to ensure that the discussion covers the length and breadth of the scope of the reliability plan and to keep in mind that the objective of the discussion is to understand the culture as it exists and to guide decisions concerning product reliability.

We'll use the Enthusiast Series to share example questions that can be used as part of the gap assessment This list of questions is only an example and not complete. In this example, we'll share three types of questions: reliability-objective questions, reliability-method questions, and reliability-organizational questions.

Reliability-objective questions for the Enthusiast project team

The gap assessment for the Enthusiast project begins with simple questions about reliability capabilities. Selected members of the Enthusiast team will be interviewed. An actual gap assessment would probably require more questions. Here are some typical questions:

1. What Enthusiast bicycle projects will need reliability strategy and planning?
2. Is there a stated Enthusiast vision for reliability? What is it? What are known gaps to this vision?
3. What has been customer feedback about the reliability of current Enthusiast products?
4. How long should Enthusiast bicycles last in service? What are known gaps in this service life?
5. Are there any known high-risk areas on the Enthusiast project? What are they?
6. How well does the reliability of current Enthusiast products compare to that of the competition?

Reliability-method questions for the Enthusiast project team

As part of the gap assessment for the Enthusiast project, it is helpful to know the capability and potential gaps for commonly used or anticipated reliability methods. For the Enthusiast project, we prepared a list of potential reliability methods to determine the capability and gaps for each method. We'll only show four methods in this example line of questioning. The list of commonly used or anticipated methods in a real program can certainly be longer. See Section 8.3.3 for reliability method categories and Appendix C for a listing of potential methods that may be useful as thought-starters.

For each of the following methods, targeted staff from the Enthusiast project will be interviewed who have knowledge about the current state and application of selected methods: setting reliability targets, system reliability modeling, HALT, and FMEA. These four questions are asked of the targeted staff for each of the selected methods:

- To what extent do you use (a selected reliability method)?
- How do you use (a selected reliability method)?
- Why do you use (a selected reliability method)?
- What examples do you have that demonstrate the application of (a selected reliability method)?

Reliability-organizational questions for the Enthusiast project team

Every reliability gap assessment needs to include as part of its scope the capability of the organization to provide the resources, training, services, and other needed support for achieving reliability objectives. Reliability-organizational questions include

- What kind of reliability support (staffing, training, and organization) is needed for the Enthusiast project? What is the current state of the Enthusiast team in terms of the needed support?
- What business processes need to be in place to support the needed reliability? What is the current state of those business processes?
- What is the scale of the product that will be developed in terms

of quantity per year and cost per item? What are known challenges to meeting the scale?

♦ Who owns reliability for Enthusiast bicycles? When does management get involved with reliability issues? Is management involvement reactive or proactive?

Advanced preparation of meaningful questions is an essential step in preparing for the gap assessment. The example for the Enthusiast project contains an incomplete list of questions. An actual reliability gap assessment will need questions from all areas of the intended scope of the reliability plan, including each stage of the product development process.

6.3.3 Preparing for a reliability gap assessment

There are several essential preparation steps to take before beginning an onsite reliability gap assessment. First, the list of questions needs to be completed so that planning can take place to get answers to the questions. Then the people or teams need to be identified so that the right people can be scheduled for interviews. Physical tours of facilities, labs, assembly plants, and other venues must be scheduled to provide a visual reality of the current scene. In addition, project and program information must be gathered together and studied to provide the backdrop for the questioning and minimize the interview times. Finally, a specific schedule for the reliability gap assessment should be agreed upon and followed.

You have to consider who needs to be consulted to get answers to each of these topics and organize the meetings and venues accordingly. Note that the gap assessment is the first step in change management: awareness. (See Section 12.5 for more information on change management.) In most cases, a reliability plan designed to close identified gaps will involve changes to how an organization currently operates.

The following comments should help you identify whom to talk with in conducting your reliability gap assessment:

- A balanced overview of the entire organization's approach to achieving reliability needs to be made.
- Whether the teams are experienced with multiple cycles of product development has to be ascertained.
- Recognize that no one person will fully understand every element of the culture or program.
- Include different levels of responsibility: technicians, engineers, managers, and senior managers
- It is acceptable for participants to pass on a question rather than guess an answer.
- Include procurement, quality, and manufacturing teams along with the design team.
- Interviews with individuals on the design team should be given more weight.
- Consider including finance, marketing, and customer support departments.

6.3.4 Conducting a reliability gap assessment

Once the preparation steps are completed, the sequence of actions for conducting a reliability gap assessment is as follows:

1. Ask questions of selected people and teams using the planned set of questions.
2. When interviewing people, pursue new topics as they come up.
3. Bring in new people or teams as needed.
4. Add to reliability gap assessment questions as needed.
5. Physically look at facilities and parts to ensure thorough answers to questions.
6. Conduct additional or clarifying interviews as needed to ensure that all questions are answered.
7. Summarize the answers to all questions.
8. Map the current state to the reliability maturity matrix. (See Section 6.3.5.)

9. Identify and list gaps from reliability gap assessment. (See Section 6.3.6.)

6.3.5 Mapping the current state to the maturity matrix

One output of the reliability gap assessment is to map the current state to the maturity matrix (see Chapter 3). Based on the information gathered about how the organization addresses reliability decisions and activities, identify within each row the maturity matrix stage that best describes the company's current status.

As covered in Chapter 3, a reliability maturity matrix is a tool for determining a company's approach to achieving product reliability performance. Where a company or project falls on the reliability maturity matrix can be assessed by analyzing the responses to questions for clues to identify the most descriptive or apt stage for each row in the matrix. A reliability maturity matrix provides a useful way to apply the tasks of a reliability plan to the unique capability of the company and helps guarantee application success.

It is important to understand where an organization maps to the maturity matrix because the gaps, and corresponding gap-closure tasks, will not be effective if they do not account for the organization's current state to implement the tasks. For example, if the company cannot perform failure mechanisms modeling, assigning a task in the reliability plan to implement physics-of-failure modeling will not have the supporting structures, processes, and understanding needed.

How do you do this exactly? The simplest way to map the gap assessment to the reliability maturity matrix is to examine each of the 11 subjects (rows) from the reliability maturity matrix and select the stage (column) that best represents the organization's capability for that subject. You can highlight the assessed verbiage from the reliability maturity matrix for each subject that is being analyzed (as done in the examples in Section 6.4).

It is often the lowest maturity by row on the maturity matrix that holds you back, and visual mapping (highlighting or circling) to the maturity matrix provides indicators to what to work on. The matrix may also reveal aspects of the organization that are doing well. These are areas of strengths and may be useful when developing a reliability plan.

Effective reliability plans must be based on the reality of where the company is today, compared to where it needs to be in the future to achieve its reliability strategic vision. Each of the gaps is input into the task selection for the reliability plan. Specifically, tasks will be selected that close the gaps and move the company toward its reliability objectives and up the stages of the maturity matrix.

6.3.6 Identify and list the gaps from the gap assessment

An analysis of the answers and discussions noted during the assessment along with a solid understanding of the reliability vision and stated reliability objectives exposes aspects of the organization that may cause the organization to not achieve its objectives. One intent of the assessment is to identify the barriers to achieving the vision and goals.

One way to identify gaps in the analysis is to consider what information is necessary to enable the achievement of the reliability goal and vision. If the current state of the organization either is unable or inadequately able to create the information, that is a gap. For example, if a key decision to launch a product is based on meeting a specific reliability performance goal, and the current organization does not have a means to estimate a design's future reliability performance, the gap is between the current inability and the future state needed to have a meaningful estimate of reliability performance.

Another type of gap may be caused by the lack of access to existing information. For example, if the customer service team conducts detailed failure analysis, yet does not share that information with the design and development team that is trying to improve an existing product, then the gap is the difference between one team needing and not having access to the very information another team is creating.

Yet another type of gap may involve the inability of current processes or methods to create adequate information that would improve the ability to achieve objectives. For example, if the current process to list and track unresolved design issues is completed by individual engineers or by separate teams, this can produce a gap (because, e.g., mechanical engineers have their own list, as do electrical engineers, etc.). The assessment can reveal that discovered problems are often not shared between teams, leading to disputes over priorities and resources. The gap is the difference between the current ineffective process and the need to prioritize and allocate resources based on salient system needs.

Consider what is new or altered as an outcome of the desired vision or objectives. To achieve a specific goal may involve using new materials, vendors, or manufacturing processes. What unanswered questions do these changes pose that need addressing? Also consider changes in markets, customer expectations, and financial goals, because these too impact the need for information that may not be currently available.

The analysis is a process to review the current state and identify those elements that hinder the ability of the team to have the right information at the right time. What information needs to be created, improved, or shared? Which processes could be updated, eliminated, or altered to improve efficiency, reduce risk, or enhance accuracy?

Gap assessment analysis is used to identify and list the gaps. The current task is not to prioritize at this time, as that is done later in the six-step process (Chapter 7). The gaps should simply be stated with enough detail to clearly convey the nature of the identified gap and its importance and ramification if not addressed. Realize, as conditions and constraints change, that some gaps will drop away and others will become apparent; consequently, this an ongoing process.

In the next chapter (Chapter 7), we will prioritize gaps and corresponding key decisions.

6.4 Examples of mapping and summarizing the reliability gap assessments

The three bicycle product lines (Pro Series, Enthusiast Series, and Intro Series) will be used to demonstrate fictitious examples of reliability gap assessments. For brevity's sake, only five gaps will be shown for each of the three product lines. In reality, there may be dozens of gaps for each product line, depending on product and process complexity and the degree of challenges necessary to meet the reliability strategic vision.

6.4.1 Pro Series: Example of mapping to the reliability maturity matrix

Based on the results of the Pro Series reliability gap assessment, the following observations can be made regarding organizational maturity:

Requirements:
- Although basic performance requirements do exist, many product requirements are inadequately defined and missing up-to-date environmental and use profiles (stage 2).
- Although the Pros Series team has considerable expertise in new bicycle technology and performance racing, there is no one familiar with reliability methods (stage 1).

Engineering:
- Test plans exist, although they are mostly dedicated to testing to specs and focused on meeting requirements (stage 2).

Feedback process:
- The Pro Series team rigorously roots out the causes of all field failures and has a sound process in place to identify and execute corrective measures (stage 3).
- There is a commitment to fixing known problems, but progress

has not yet been made in taking the reliability engineering actions necessary to prevent problems (stage 3).

Management:

- Although there has not been adequate reliability planning and management in previous programs, the Pro Series team is committed to putting more focus on reliability management (stage 3).
- Currently, reliability is hidden in manufacturing or engineering departments. Emphasis has been on initial product functionality while conducting routine product testing (stage 2).

Based on these and other observations, the stage of the maturity matrix that best represents the current status of the Pro Series team is stage 2: awakening (see Table 6.1).

		Stage 1: Uncertainty	Stage 2: Awakening	Stage 3: Enlightenment	Stage 4: Wisdom	Stage 5: Certainty
Requirements	Requirements & Planning	Informal or nonexistent.	Basic requirements based on customer requirements or standards. Plans have required activities.	Requirements include environment and use profiles. Some apportionment done. Plans have more details with regular reviews.	Plans are tailored for each project and projected risks. Use of distributions for environmental and use conditions.	Contingency planning occurs. Decisions based on business or market considerations. Part of strategic business plan.
	Training & Development	Informally available to some, if requested.	Select individuals trained in concepts and data analysis. Available training for design engineers.	Training for engineering community on key reliability-related processes. Managers trained on reliability and lifecycle impact.	Reliability and statistics courses tailored for design and manufacturing engineers. Senior managers trained on reliability impact on business.	New technologies and reliability tools tracked and training adjusted to accommodate. Reliability training supported.
Engineering	Analysis	Nonexistent or based on manufacturing issues.	Point estimates and reliance on handbook parts count methods. Basic identification and listing of failure modes and impact.	Formal use of FMEA. Field data analysis of similar products used to adjust predictions. Design changes cause reevaluation of product reliability.	Predictions are expressed as distributions and include confidence limits. Environmental and use conditions used for simulation and testing.	Lifecycle cost considered during design. Stress and damage models created and used. Extensive risk analysis performed as needed.
	Testing	Primarily functional.	Generic test plan exists with reliability testing only to meet customer or standards specifications.	Detailed reliability test plan with sample size and confidence limits. Results used for design changes and vendor evaluations.	Accelerated tests and supporting models used. Testing to failure or destruct limits conducted.	Test results used to update component stress and damage models. New technologies characterized.
	Supply Chain Management	Selection based on function & price.	Approved vendor list maintained. Audits based on issues and critical parts. Qualification primarily based on datasheets.	Assessments and audit results used to update AVL. Field data collected and failure analysis performed related to specific vendors.	Vendor selection includes analysis of vendor's reliability data. Suppliers conduct assessments and audit of their suppliers.	Changes in environment, use profile, or design trigger vendor reliability assessment. Part parameters and reliability monitored for stability.
Feedback Process	Failure Data Tracking & Analysis	May address function testing failures.	Pareto analysis of field return and internal testing. Failure analysis relies on vendor support.	Root cause analysis used to update AVL and prediction models. Summary of analysis results disseminated.	Focus is on failure mechanisms. Failure distribution models updated based on failure data.	Customer satisfaction relationship to failures is understood. Prognostic tools used to forestall failure.
	Validation & Verification	Informal and based on individuals rather than process.	Basic verification that plans are followed. Field failure data regularly reported.	Supplier agreements around reliability monitored. Failure modes regularly monitored.	Internal reviews of processes and tools. Failure mechanisms monitored and used to update models and test methods.	Reliability predictions match observed field reliability.
	Improvement	Nonexistent or informal.	Design & process change processes followed, corrective action taken.	Effectiveness of corrective actions tracked over time. Identified failure modes addressed in other products. Improvement opportunities identified as stresses and use profiles change.	Identified failure mechanisms addressed in all products. Advanced modeling techniques explored and adopted. Formal and effective lessons-learned process exists.	New technologies evaluated and adopted to improve reliability. Design rules updated based on field failure analysis.
Management	Understanding & Attitude	No comprehension of reliability as a management tool. Tendency to blame engineering for reliability problems.	Recognizing that reliability management may be of value but not willing to provide money or time to make it happen.	Still learning more about reliability management. Becoming supportive and helpful.	Full participation. Understanding of absolutes of reliability management. Recognizing their personal role in continuing emphasis.	Consider reliability management an essential part of the company system.
	Status	Reliability is hidden in manufacturing or engineering departments. Reliability testing probably not done. Emphasis on product functionality.	A stronger reliability leader appointed, yet main emphasis is still on an audit of initial product functionality. Reliability testing still not performed.	Reliability manager reports to top management, while in management of division.	Reliability manager is an officer of the company, reporting effective status and devising preventive action and involved with consumer affairs.	Reliability manager is on the board of directors. Prevention is the main concern. Reliability is a thought leads.
	Cost of Unreliability	Not done other than anecdotally.	Direct warranty expenses only.	Warranty, corrective action materials, and engineering costs monitored.	Customer and lifecycle unreliability costs identified and tracked.	Lifecycle cost reduction done through product reliability improvements.
	Prevailing Sentiment	"We don't know why we have problems with reliability."	"Is it absolutely necessary to always have problems with reliability?"	"Through commitment and improvement, we are identifying and resolving our problems."	"Failure prevention is a routine part of our operation."	"We know why we do not have problems with reliability."

Table 6.1. Reliability maturity matrix for the Pro Series.

(Full size matrix available at accendoreliability.go/pre)

6.4.2 Pro Series: Partial example of gaps

Strength. The Pro Series team rigorously roots out the causes of all field failures and has a good process in place to identify and execute corrections.

Strength. The Pro Series team is committed to putting more focus on reliability management.

Gap. Current Pro Series test procedures use mostly pass–fail tests, and the conditions-of-use profiles for professional bicyclists are outdated. Given the need for extreme durability under all road conditions, the test regimes need to be changed to test to failure. The Pro Series team needs to research and incorporate conditions-of-use profiles consistent with anticipated extreme conditions.

Gap. With the low volumes available for testing, more emphasis needs to be placed on proper design guidelines for the Pro Series, bringing together the latest materials, technology, and craftsmanship. Existing design guides are missing the latest materials and technology.

Gap. Ease of maintenance and servicing is essential to the racing customers; however, there are no preventive maintenance plans in place on current systems. Full preventative maintenance plans must be developed.

Gap. A review of the reliability organization reveals that no one is well trained on reliability methods. According to the reliability strategic vision, the entire Pro Series team must be trained on reliability and availability.

Gap. A review of past Pro Series FMEAs indicates that they were done poorly, with little to no follow-up or risk reduction. To design-in reliability, a system FMEA and critical component FMEAs will need to be done. None of the Pro Series team is adequately trained in FMEA procedure.

6.4.3 Enthusiast Series: Examples of mapping to the reliability maturity matrix

Based on the results of the Enthusiast Series reliability gap assessment, the following observations can be made regarding organizational maturity:

Requirements:

- ♦ Environment and use profiles exist; however, they do not include distributions for use profiles and need to be updated (stage 3).

Engineering:

- ♦ There is an approved list of vendors. The criteria include field data analysis, performance capability, and audits (stage 3).
- ♦ There are some analytical tools being used by the Enthusiast Series team, such as Weibull analysis of test and field data, use of manufacturing control charts, and reliability prediction based on parts count; a review of FMEAs reveals poor quality (stage 2).

Feedback Process:

- ♦ The Enthusiast Series team has a workable corrective action process and tracks most execution changes; however, for the most prevalent field failure mode, there is no planned resolution to the problem (stage 2).

		Stage 1: Uncertainty	Stage 2: Awakening	Stage 3: Enlightenment	Stage 4: Wisdom	Stage 5: Certainty
Requirements	Requirements & Planning	Informal or nonexistent.	Basic requirements based on customer requirements or standards. Plans have required activities.	Requirements include environment and use profiles. Some apportionment done. Plans have more details with regular reviews.	Plans are tailored for each project and projected risks. Use of distributions for environmental and use conditions.	Contingency planning occurs. Decisions based on business or market considerations. Part of strategic business plan.
	Training & Development	Informally available to some, if requested.	Select individuals trained in concepts and data analysis. Available training for design engineers.	Training for engineering community on key reliability-related processes. Managers trained on reliability and lifecycle impact.	Reliability and statistics courses tailored for design and manufacturing engineers. Senior managers trained on reliability impact on business.	New technologies and reliability tools tracked and training adjusted to accommodate. Reliability training supported.
Engineering	Analysis	Nonexistent or based on manufacturing issues.	Point estimates and reliance on handbook parts count methods. Basic identification and listing of failure modes and impact.	Formal use of FMEA. Field data analysis of similar products used to adjust predictions. Design changes cause reevaluation of product reliability.	Predictions are expressed as distributions and include confidence limits. Environmental and use conditions used for simulation and testing.	Lifecycle cost considered during design. Stress and damage models created and used. Extensive risk analysis performed as needed.
	Testing	Primarily functional.	Generic test plan made with reliability testing only to meet customer or standards specifications.	Detailed reliability test plan with sample size and confidence limits. Results used for design changes and vendor evaluations.	Accelerated tests and supporting models used. Testing to failure or destruct limits conducted.	Test results used to update component stress and damage models. New technologies characterized.
	Supply Chain Management	Selection based on function & price.	Approved vendor list maintained. Audits based on issues and critical parts. Qualification primarily based on datasheets.	Assessments and audit results used to update AVL. Field data collected and failure analysis performed related to specific vendors.	Vendor selection includes analysis of vendor's reliability data. Suppliers conduct assessments and audit of their suppliers.	Changes in environment, use profile, or design trigger vendor reliability assessment. Part parameters and reliability monitored for stability.
Feedback Process	Failure Data Tracking & Analysis	May address function testing failures.	Pareto analysis of field return and internal testing. Failure analysis relies on vendor support.	Root cause analysis used to update AVL and prediction models. Summary of analysis results disseminated.	Focus is on failure mechanisms. Failure distribution models updated based on failure data.	Customer satisfaction relationship to failures is understood. Prognostic tools used to forestall failure.
	Validation & Verification	Informal and based on individuals rather than process.	Basic verification that plans are followed. Field failure data regularly reported.	Supplier agreements around reliability monitored. Failure modes regularly monitored.	Internal reviews of processes and tools. Failure mechanisms monitored and used to update models and test methods.	Reliability predictions match observed field reliability.
	Improvement	Nonexistent or informal.	Design & process change processes followed, corrective action taken.	Effectiveness of corrective actions tracked over time. Identified failure modes addressed in other products. Improvement opportunities identified as stresses and use profiles change.	Identified failure mechanisms addressed in all products. Advanced modeling techniques explored and adopted. Formal and effective lessons learned process exists.	New technologies evaluated and adopted to improve reliability. Design rules updated based on field failure analysis.
Management	Understanding & Attitude	No comprehension of reliability as a management tool. Tendency to blame engineering for reliability problems.	Recognizing that reliability management may be of value but not willing to provide money or time to make it happen.	Still learning more about reliability management. Becoming supportive and helpful.	Full participation. Understanding of absolutes of reliability management. Recognizing their personal role in continuing emphasis.	Consider reliability management an essential part of the company system.
	Status	Reliability is hidden in manufacturing or engineering departments. Reliability testing probably not done. Emphasis on product functionality.	A stronger reliability leader appointed, yet main emphasis is still on an audit of initial product functionality. Reliability testing still not performed.	Reliability manager reports to top management, with role in management of division.	Reliability manager is an officer of the company, reporting effective status and devising preventive action and involved with consumer affairs.	Reliability manager is on the board of directors. Prevention is the main concern. Reliability is a thought leader.
	Cost of Unreliability	Not done other than anecdotally.	Direct warranty expenses only.	Warranty, corrective action materials, and engineering costs monitored.	Customer and lifecycle unreliability costs identified and tracked.	Lifecycle cost reduction done through product reliability improvements.
	Prevailing Sentiment	"We don't know why we have problems with reliability."	"Is it absolutely necessary to always have problems with reliability?"	"Through commitment and improvement, we are identifying and resolving our problems."	"Failure prevention is a routine part of our operation."	"We know why we do not have problems with reliability."

Table 6.2. Reliability maturity matrix for the Enthusiast Series.

(Full size matrix available at accendoreliability.go/pre)

Management:

- A Reliability Manager has been appointed who reports to the Engineering Director; plans are in place to train the Enthusiast Series team on reliability methods (stage 3).

Based on these and other observations, the stage of the maturity matrix that best represents the current status of the Enthusiast Series team is somewhere between stage 2: awakening and stage 3: enlightenment (see Table 6.2).

6.4.4 Enthusiast Series: Partial examples of gaps

Strength. There is an approved list of vendors based on criteria including field data analysis, performance capability, and audits.

Strength. Plans are in place to train the Enthusiast Series team on reliability statistical methods.

Gap. Conditions of use profiles are used to drive design and testing; however, they do not represent the most severe conditions anticipated.

Gap. Current supplier selection is based mostly on cost parameters, with some consideration of supplier reliability and performance capability.

Gap. The most prevalent field failure mode for Enthusiast Series bicycles is the chain slipping off the derailleur, and there is no fix for this problem. No effort has been made to make the derailleur subsystem more robust in future designs.

Gap. A review of design guidelines for Enthusiast Series design engineers shows no mention of tasks or activities that improve reliability.

Gap. Although Enthusiast Series customers will be expected to maintain their bicycles at regular maintenance intervals, the existing published preventive maintenance procedures are inadequate.

6.4.5 Intro Series: Examples of mapping to the reliability maturity matrix

Based on the results of the Intro Series reliability gap assessment, the following observations can be made regarding organizational maturity:

Requirements:

- There are no reliability requirements for the project, and performance requirements are very informal (stage 1).

- There are no reliability resources or training within the Intro Series team, and there are no plans to leverage reliability resources from other teams (stage 1).

Engineering:

- Supplier selection is done based primarily on price and does not consider the supplier's safety or reliability capability (stage 1).

- Testing is primarily functional and based on pass–fail criteria, with the objective of meeting standards; there is no reliability testing done by the Intro Series team (stage 2).

Feedback Process:

- Design and process change processes are followed by the management team; corrective action processes are somewhat defined (stage 2).

- Verification plans are not formally defined; no process exists to track failures to resolution (stage 1).

- Failures are identified during testing internally with no root cause analysis performed; vendors are pressured to resolve problems (stage 2).

Management:

- The Intro Series team has no real knowledge of reliability methods, and there are no plans to develop reliability expertise (stage 1).

		Stage 1: Uncertainty	Stage 2: Awakening	Stage 3: Enlightenment	Stage 4: Wisdom	Stage 5: Certainty
Requirements	Requirements & Planning	Informal or nonexistent.	Basic requirements based on customer requirements or standards. Plans have required activities.	Requirements include environment and use profiles. Some apportionment done. Plans have more details with regular reviews.	Plans are tailored for each project and projected risks. Use of distributions for environmental and use conditions.	Contingency planning occurs. Decisions based on business or market considerations. Part of strategic business plan.
	Training & Development	Informally available to some, if requested.	Select individuals trained in concepts and data analysis. Available training for design engineers.	Training for engineering community on key reliability-related processes. Managers trained on reliability and lifecycle impact.	Reliability and statistics courses tailored for design and manufacturing engineers. Senior managers trained on reliability impact on business.	New technologies and reliability tools tracked and training adjusted to accommodate. Reliability training supported.
Engineering	Analysis	Nonexistent or based on manufacturing issues.	Point estimates and reliance on handbook parts count methods. Basic identification and listing of failure modes and impact.	Formal use of FMEA. Field data analysis of similar products used to adjust predictions. Design changes cause reevaluation of product reliability.	Predictions are expressed as distributions and include confidence limits. Environmental and use conditions used for simulation and testing.	Lifecycle cost considered during design. Stress and damage models created and used. Extensive risk analysis performed as needed.
	Testing	Primarily functional.	Generic test plan exists with reliability testing only to meet customer or standards specifications.	Detailed reliability test plan with sample size and confidence limits. Results used for design changes and vendor evaluations.	Accelerated tests and supporting models used. Testing to failure or destruct limits conducted.	Test results used to update component stress and damage models. New technologies characterized.
	Supply Chain Management	Selection based on function & price.	Approved vendor list maintained. Audits based on issues and critical parts. Qualification primarily based on datasheets.	Assessments and audit results used to update AVL. Field data collected and failure analysis performed related to specific vendors.	Vendor selection includes analysis of vendor's reliability data. Suppliers conduct assessments and audit of their suppliers.	Changes in environment, use profile, or design trigger vendor reliability assessment. Part parameters and reliability monitored for stability.
Feedback Process	Failure Data Tracking & Analysis	May address function testing failures.	Pareto analysis used from internal testing. Failure analysis relies on vendor support.	Root cause analysis used to update AVL and prediction models. Summary of analysis results disseminated.	Focus is on failure mechanisms. Failure distribution models updated based on failure data.	Customer satisfaction relationship to failures is understood. Prognostic tools used to forecast failure.
	Validation & Verification	Informal and based on individuals rather than process.	Basic verification that plans are followed. Field failure data regularly reported.	Supplier agreements around reliability monitored. Failure modes regularly monitored.	Internal reviews of processes and tools. Failure mechanisms monitored and used to update models and test methods.	Reliability predictions match observed field reliability.
	Improvement	Nonexistent or informal.	Design & process change processes followed, corrective action taken.	Effectiveness of corrective actions tracked over time. Identified failure modes addressed in other products. Improvement opportunities identified as stresses and use profiles change.	Identified failure mechanisms addressed in all products. Advanced modeling techniques explored and adopted. Formal and effective lessons-learned process exists.	New technologies evaluated and adopted to improve reliability. Design rules updated based on field failure analysis.
Management	Understanding & Attitude	No comprehension of reliability as a management tool. Tendency to blame engineering for reliability problems.	Recognizing that reliability management may be of value but not willing to provide money or time to make it happen.	Still learning more about reliability management. Becoming supportive and helpful.	Full participation. Understanding of absolutes of reliability management. Recognizing their personal role in continuing emphasis.	Consider reliability management an essential part of the company system.
	Status	Reliability is hidden in manufacturing or engineering departments. Reliability testing probably not done. Emphasis on product functionality.	A stronger reliability leader appointed, yet main emphasis is still on an audit of initial product functionality. Reliability testing still not performed.	Reliability manager reports to top management, with role in management of division.	Reliability manager is an officer of the company, reporting effective status and devising preventive action and involved with consumer affairs.	Reliability manager is on the board of directors. Prevention is the main concern. Reliability is a thought leader.
	Cost of Unreliability	Not done other than anecdotally.	Direct warranty expenses only.	Warranty, corrective action materials, and engineering costs monitored.	Customer and lifecycle unreliability costs identified and tracked.	Lifecycle cost reduction done through product reliability improvements.
	Prevailing Sentiment	"We don't know why we have problems with reliability."	"Is it absolutely necessary to always have problems with reliability?"	"Through commitment and improvement, we are identifying and resolving our problems."	"Failure prevention is a routine part of our operation."	"We know why we do not have problems with reliability."

Table 6.3. Reliability maturity matrix for the Intro Series.
(Full size matrix available at accendoreliability.go/pre)

Based on these and other observations, the stage of the maturity matrix that best represents the current status of the Intro Series team is somewhere between stage 1: uncertainty and stage 2: awakening (see Table 6.3).

6.4.6 Intro Series: Partial examples of gaps

Strength. The Intro Series management has recognized its lack of reliability resources and is planning on using reliability expertise from the other bicycle projects.

Strength. Design and process change processes are followed by the management team. Corrective action processes are somewhat defined.

Gap. Verification plans are not formally defined; no process exists to track failures to resolution.

Gap. System safety is stated to be a high priority; however, the Intro Series team did not know how they would be ensuring system safety.

Gap. The supplier selected for brake pads has a track record of safety recalls. No system is in place to consider safety in the supplier selection process.

103

Gap. Intro Series customers will have minimal maintenance requirements. However, three bicycle subsystems for the Intro Series require regular maintenance to prevent premature failures.

Gap. Intro Series customers want the bicycle to be easy to use. None of the current test procedures address the "ease of use" of the various bicycle subsystems.

6.5 Summary

One output of the assessment is an understanding of the current reliability maturity of the organization. Documenting the maturity by row of the matrix provides an effective communication tool and provides information necessary to create an effective reliability plan.

The differences between an organization's current and needed capabilities and information define the gaps that must be overcome to achieve the reliability vision and goals. This close look at how a team uses the available information to make decisions impacting reliability may reveal many potential areas for improvement.

In the next chapter, we will further refine what is necessary to achieve the goals and vision by focusing on the decisions that define reliability performance. The next step is to narrow down the list of potential improvements to those that are most essential to achieving the reliability goals and vision.

Chapter 7

IDENTIFYING RELIABILITY-RELATED DECISIONS

Management is doing things right; leadership is doing the right things.
Peter Drucker

In this chapter. We continue the process of gathering information to build a reliability plan. This step entails identifying what information needs to be generated via the tasks in the plan. The idea is to gather all decisions that would benefit from reliability-related information, then narrow down the list to those most important to achieving the desired reliability objectives.

7.1 What is meant by reliability-related decisions?

For a given project or program, there are many tools, tests, or analyses to support achieving product reliability objectives. Yet, which minimum set of such activities will assist the program or project efficiently and effectively? What should be the focus of the reliability plan such that the results provide the most value?

The intent of this chapter is to provide a means to focus the reliability plan development on the vital few key decisions and to build on existing capabilities. By understanding the nature of those key decisions, as well as the risk posed by the identified gaps, and by employing the capability of the team, you can facilitate the selection of the right set of reliability tools for the plan.

7.2 Where does identifying reliability-related decisions fit into the six-step process?

Identifying reliability-related decisions is the third of the six steps taken to achieve high reliability. The reliability strategic vision and reliability gap assessment are inputs to identifying key program decisions.

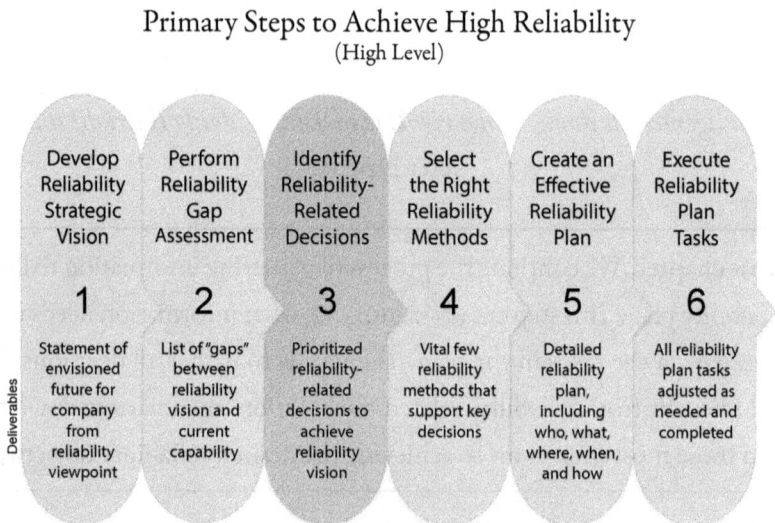

Primary Steps to Achieve High Reliability
(High Level)

	Develop Reliability Strategic Vision	Perform Reliability Gap Assessment	Identify Reliability-Related Decisions	Select the Right Reliability Methods	Create an Effective Reliability Plan	Execute Reliability Plan Tasks
	1	2	3	4	5	6
Deliverables	Statement of envisioned future for company from reliability viewpoint	List of "gaps" between reliability vision and current capability	Prioritized reliability-related decisions to achieve reliability vision	Vital few reliability methods that support key decisions	Detailed reliability plan, including who, what, where, when, and how	All reliability plan tasks adjusted as needed and completed

7.3 Gaps and decisions as input to the reliability plan

Identifying the gaps and capabilities of an organization enables the creation of a list of opportunities for improvement and a list of capabilities to maintain. Each reliability-related capability within an organization supports one or more decisions by providing insights, information, or results. The key capabilities and associated decisions are those to carry forward to the detailed plan development. These capabilities impede or support the organization's ability to achieve the desired objectives and vision.

Gap assessment results in a list of identified differences between the desired reliability vision and current capabilities. Closing a gap often requires changing or adding a new capability. It may require providing training, acquiring new equipment, altering business processes, or improving communication. Identifying gaps typically leads to a long list of potential gaps; not all of them, when closed, provide the same value. Some will have a large impact, while others will have a minor impact on the team's ability to create a reliable product. Prioritizing which gaps to address in the reliability plan is essential to crafting the plan.

Another aspect of the gap assessment is the identification of areas that are doing well, that is, areas with no gaps. These activities, tools, processes, capabilities, etc. support the team's ability to achieve the organization's reliability vision and project goals.

Focusing only on closing gaps will help a team make improvements, yet it is also important to include what the team already does well. If a current capability would provide sufficient information for a key decision (as will be discussed below) or would enhance the closing of a gap, then it should find a place in the reliability plan. If what is being done well now will add value, you should consider including that task or tool in the reliability plan.

The intent of considering gaps is to prioritize the associated decisions impacted by the existence of the gap(s) and address closing the gap(s) with the reliability plan activities and results.

In addition to addressing gaps, keep in mind that the reliability of an item is established one decision at a time. Members of the team, often not the reliability engineers or managers, determine the type of materials, identify which vendors to engage, and address thousands of other details that make up a product's design and resulting reliability performance.

It is important to recognize that a reliability plan will not be able to address every decision. It can, however, influence and improve the key decisions as well as provide guidelines and best practices to enable better decision-making across the organization. The key decisions are those that have a significant impact on the resulting reliability performance of the item. They are the decisions that benefit from clear information on potential risks and reliability performance.

Many organizations that develop products have a product life-cycle document that outlines the phases of development and the key elements to address within each phase. The product life cycle may include concept, development, production, and field support phases. Each product life cycle applies to the unique needs of the organization and may include a different number of phases. The product life cycle may provide a general guideline for the development process or a detailed checklist for each milestone. We mention the product life cycle here as it may provide useful information to help identify key milestone decisions the team will have to address.

The intent of considering decisions is to prioritize those key decisions that would benefit from information generated by the output of specific reliability activities. There are three considerations when prioritizing what to address in a reliability plan: gaps that may impede achieving the vision, strengths that add value, and key decisions that would benefit from adequate reliability information.

Gaps and decisions are not the only factors to consider. Factors such as timeline, budget, regulatory or customer requirements, and market expectations also play a role in crafting a reliability plan. For any given industry or situation, the careful review of gaps and decisions provides a solid foundation for the plan's success. The other factors serve to shape the selection of specific activities for inclusion in the plan.

7.4 Why is it important to identify key decisions?

When asked to conduct a design FMEA, one question to pose is "Why?" Why are we investing the time and resources to conduct this analysis? One common reason is to identify and prioritize work to reduce potential safety and reliability performance problems. The underlying decision may involve how to allocate resources to maximize reliability improvements with the current design, which is likely an important set of decisions for the project manager in charge of setting priorities and allocating resources for the project.

Again, when asked to run a test or analyze field data, you need to know why. Who needs the results and for what reason? What decision will the results of the task influence? Focus should be placed on identifying decisions that benefit from the output of reliability activities. A check step then becomes asking why each planned activity is needed to ensure it is still connected to a key decision.

Key decisions are those that impact the safety and performance of the product or system. The decision may include factors such as time to market, cost, functional performance, and reliability performance. For example, should the design use a harder material with longer durability yet with a higher cost, or should we use the softer material that wears quicker at a lower cost? Knowing the specifics of the two materials' reliability performance would significantly influence the selection of the material for the design.

Decisions occur at all levels of the organization. For each project, there may be a few strategic decisions and hundreds of deliberate decisions that directly define the item's reliability performance. These decisions often have an obvious relationship to reliability performance. There are also thousands of decisions made that do not obviously impact reliability, yet they do. Many decisions are made in a routine manner without considering the impact on reliability. These decisions may include component placement or design drawing tolerances; these may have a major impact individually and collectively.

Not all decisions are the same. Many decisions have little overall impact on the reliability performance. For example, when sourcing a component, two vendors may offer very similar solutions, technology, field performance, costs, risks, etc. Choosing one or the other vendor's component may have very little impact on the reliability of the system. However, some decisions significantly impact the eventual reliability performance. For example, someone will have to decide whether the product design is complete and likely to meet the reliability objectives. Another example might be vendor selection in which there are significant differences in each vendor's technology, field performance, costs, risks, etc.

Furthermore, not all decisions require the same detail of reliability information to improve the decision's proper consideration of reliability impact. A pending decision with a recognized potential major impact may require an extensive effort to gather or create the necessary information to inform the decision makers. Other decisions may only require a quick test or a simple design guideline.

Another group of decisions that do have an impact comprises those done without any conscious consideration of the potential reliability impact. These are difficult to identify, as even the decision maker does not realize the relationship between the decision outcome on reliability performance. An example might be a change in supplier as a cost-cutting measure, without understanding the reliability impact of the change in supplier. As the organization's reliability culture matures, decision makers learn to always consider the reliability impact of all decisions.

7.5 Gathering potential key decisions

At the start and during the entire product life cycle, decisions are made. Many would benefit from relevant reliability-related information. Potential key decisions are those the stakeholders and team members discuss and actively seek information to inform their decisions. When prioritizing decisions to support, priority should be placed on those having the most impact on the reliability performance of the item.

Some decisions have already been made and become input to the plan as they will still require the allocation of resources to address or implement. Mandates and requirements reflect the result of decisions made by authorities able to impose directives to include within the reliability plan. Mandates may be imposed by an authority within or outside your organization. For example, the Director of Engineering is requiring electrical engineering work to include adherence to a new derating guideline. A key customer is requiring the use of a specific coating or finish with which the organization does not have prior experience. Requirements may arise from government or regulatory bodies. For example, a new requirement for bicycles sold shall have a nonremovable serial number or embedded identification device.

Strengths identified during the gap assessment are another element to consider along with key decisions. Strengths provide an existing capability to inform decisions that still need to be addressed in the plan. In addition, leveraging areas of strength may provide an efficient means to close a gap.

The reliability plan uses this prioritized list of key decisions, mandates, and requirements, as well as strengths, as part of the input to building a plan.

One needs to gather the range of upcoming decisions that may benefit from information generated by the reliability plan's activities and tasks. This is not a one-time task. As the project moves along the product development process, one needs to continue to identify decisions that would benefit from reliability information.

Reviewing the gap assessment results and the product life-cycle guidance provides a start to gathering key decisions. Various stakeholders may make requests for information or specific activities to include in the plan. These requests have a pending decision.

Interviews should be conducted with stakeholders to question them about their upcoming decisions and what specific reliability information would benefit the outcomes of those decisions. Another way to learn about key decisions is during development team meetings. Even simply listening during casual encounters or discussions can be beneficial.

We next focus on gap assessment and associated decisions and provide some examples based on the Enthusiast Series bicycle. Many of these key decisions may have first been identified during the gap assessment, which is fine. This step also includes the process to once again check with all available resources and stakeholders to ensure that all potential key decisions are considered. The following examines each potential source of reliability decisions that may need addressing in the plan using the Enthusiast Series example..

Gap assessment results. The list of gaps identified during the gap assessment is a great source of potential key decisions. Gaps are identified disparities between what is necessary to create a highly reliable product and current capabilities. Consider each gap and identify the associated decision and add those decisions to the list to prioritize.

Gaps exist because someone within the organization will require information to make a decision, and the current capabilities cannot provide adequate reliability information to support making the right decision.

For example, consider the five identified gaps from the prior analysis of the Enthusiast Series (from Chapter 6) and note the associated decisions based on our understanding of the gaps and why those gaps exist. This is summarized in Table 7.1.

Gap	Decision
Conditions of use profiles are used to drive design and testing; however, they do not represent the most severe conditions that are anticipated.	Engineering (transaction-level) decisions related to design margin and safety factors are based on expected use profiles and environmental conditions
Current supplier selection is based mostly on cost parameters, with insufficient consideration of supplier reliability and performance capability.	Strategic, operational, and transaction-level decisions are made on which vendors to engage for the project.
The most prevalent field failure mode for Enthusiast Series bicycles is the chain slipping off the derailleur; there is no fix for this problem. No current effort is being made to make the derailleur subsystem more robust in future designs.	Determine the root cause analysis and potential solutions to the chain slipping off the derailleur.
A review of design guidelines for Enthusiast Series design engineers reveals no mention of tasks or activities that improve reliability	Establish how the design guidelines for the design engineers should change to include expected reliability improvement activities.
Enthusiast Series customers will be expected to maintain their bicycles at regular intervals, but the existing published preventive maintenance procedures are inadequate.	Decide how the published preventative maintenance procedures should change to become an adequate resource for Enthusiast Series customers.

Table 7.1. Gaps and related decisions for the Enthusiast Series example.

Here we are using imaginary gaps and associated decisions to provide an example. In practice, each gap identified will have one or more associated decisions. You can identify those decisions during the gap assessment, in discussion with key stakeholders, or from your knowledge of the needs of the team to make better decisions based on adequate reliability information.

In addition, you should list and consider the identified strengths from the gap assessment. Building on strength may facilitate closing a gap or informing a key decision. Here are two strengths, as an example, of the Enthusiast Series (from Chapter 6):

1. There is an approved list of vendors based on criteria, including field data analysis, performance capability, and audits.

2. Plans are in place to train the Enthusiast Series team on reliability statistical methods.

During the entire process of collecting reliability-related decisions, add side notes of potential opportunities or connections. As an example, the strength of the approved vendor list may provide an opportunity to close the gap concerning the reliability consideration with vendor selection by incorporating information in the existing business processes for vendor selection. In this example, closing the gap may require a relatively minor investment or change of a business process or policy to make it happen, thus making closing the gap relatively easy.

Product life-cycle guideline. The product life-cycle document may provide a list of requirements for information or activities for each life-cycle milestone. The milestones reflect the necessary decision to continue working in the current phase or to proceed to the next phase of the life cycle. At the milestone between the end of design and development and before starting production, the team faces the key decision: Does the design meet the reliability objectives?

Requests. Requests may arise from anywhere in the organization. Team members may approach with specific requests for information, research, or experiments. These requests have a motivating decision behind them, so a quick conversation to understand the request fully and the associated decision allows capturing the decisions for consideration when building the reliability plan. For example, the mechanical engineer developing the frame is considering a new alloy and is not sure of the appropriate welding technique to ensure that the frame structure has the ability to meet the reliability objectives. The underlying decision is to decide which welding technique to use in manufacturing the bicycle frame.

Interviews. Interviews with key stakeholders will often occur as a normal activity during the various steps in achieving high reliability. They can occur while developing a reliability strategic vision or during the gap assessment. Interviews with stakeholders can also be helpful to discuss what they need to know as they face upcoming decisions and provide valuable input to the reliability plan. Discussing design challenges, areas with uncertainty concerning reliability performance, and the decisions that a person is facing does two things: First, you add to the list of decisions that the plan's activities may assist in or influence. Second, the person being interviewed learns what the reliability plan may provide to benefit him or her in creating a reliable product. For example, during an interview with the procurement manager, she mentions the need to change the project's tire supplier. A brief discussion reveals the underlying key decision: What is the reliability risk associated with changing tire suppliers for this project?

You should plan on talking to as many of the stakeholders as possible, including design and manufacturing engineers, procurement and finance personnel supporting the project, the project manager or lead, senior-level managers, and customers. Try to ascertain what reliability information or analysis would benefit them as they face making decisions that will impact the product's reliability performance. For example, the engineer designing the new shipping containers for the project mentioned that his goal is to reduce shipping-induced damage and would like a breakdown of past out-of-box failure reports. Learning about the prior salient failures caused by shipping-induced damage allows us to focus on the major prior problems.

Team discussions and meetings. Team discussions and meetings often include a summary of achievements, challenges, and deadlines. New information that becomes available may require a change in timeline, thus impacting the timeline for other key decisions. Changes in component or material availability or costs may entail exploring alternative design solutions. Achieving design objectives may require using an unproven manufacturing technique with an unknown impact on product reliability. For example, the procurement manager informed the team that the lead time for the new prototype sprockets has increased from four to eight weeks. What changes to the prototype testing schedule will we have to make given the four-week increase in sprockets lead time?

Listening. Listening during interviews, team meetings, and any encounter with team members allows you to identify the topics of interest to your peers and management team. When you hear about a challenge, consider what reliability information or activity may assist in understanding and overcoming the challenge. When you hear about changes to product requirements or program milestones, consider how those changes may benefit reliability information or how those changes impact when that reliability information is needed for the decision makers involved.

Understanding the full range of decisions facing the team allows the reliability plan to create outcomes that influence decisions, support the creation of a reliable product, and enable the entire team to meet reliability and business objectives. Remember that the plan is in service to the team; it is not a set of hurdles for the team to overcome.

7.6 Ranking potential key decisions

The challenge after a gap assessment and gathering potential key decisions is narrowing down the list of opportunities for improvement. Each decision identified may provide a way to improve the ability of the team to create a highly reliable product. However, not every decision poses the same risk to achieving the reliability objectives or the same opportunity for improvement.

7.6.1 Assessing the risks posed to the project

With a small team familiar with the reliability vision, objectives, and gap assessment results, it is best to rank order and score each decision on a scale from low to high risk to the project successfully achieving the reliability objectives. The idea is to identify the relative risk posed by each decision if that decision is not adequately informed for this particular project.

Consider a couple of examples from the list of key decisions:

One listed decision from the gap assessment is the set of decisions on which vendors to engage for the project. With its knowledge of the organization and the gap assessment results, the team may know that the most common field failures for recent product launches were the result of vendor process changes. Given the magnitude of recent vendor-related field problems, not improving the entire team's ability to select vendors with adequate reliability consideration is a "high" risk to the current project.

Another example is the upcoming shipping container design decisions related to reducing shipping-related damage. The team compares the risk posed by the shipping-related damage to other risks or opportunities and determines that the relative risk ranking is lower than the risk posed by the current supplier selection process, thus assigning this decision to "low" risk. The team continues to evaluate each potential key decision and assign either a low or high risk to the project. The team may also find it useful to list the decisions in relative rank order from low to high or to use a suitable scale to assign risks to each decision. Here we are using a simple scale of low or high risk to the project for the purpose of the example.

7.6.2 Assessing the difficulty of issues

Another factor to consider is that some decisions are easier to inform than others. Some decisions may take little investment to act upon by leveraging an activity the team already does well, whereas other decisions may require systemic changes to how the organization conducts product development. Reviewing the resulting maturity matrix from the gap assessment process will assist in identifying areas that may impede or support the team's ability to inform a decision properly. This time, we will use a small team that is familiar with the gap assessment, the gathered list of decisions, and the range of potential reliability activities that may be useful. The objective is to sort the decisions on a scale related to the difficulty, using a scale from easy (little investment or effort) to hard (implying a large investment and effort). This process balances the needs of the decision maker with the ability of one or more reliability methods to provide adequate information, thus reducing the decision's risk posed to the project.

Considering the same two decisions as above, as an example, let's assign difficulty to each decision: The first example from the gap assessment comprises the strategic-, operational-, and transaction-level decisions on which vendors to engage for the project. The team considers the strength of an existing approved vendor list and the rich set of field failure analyses that is available. Adding vendor-specific field reliability performance to the approved vendor list while making minor adjustments to vendor selection policy should be relatively quick and easy to accomplish. Therefore, the team assigns a "low" difficulty score to this decision.

The second example from an interview is the upcoming shipping container design decisions that should reduce shipping-related damage. The team knows the design of past shipping containers needs improvement, and they know that the engineer currently assigned to address the shipping container problem may require the assistance of a shipping container consultant to devise significant and cost-effective improvements. Therefore, the team assigns a difficulty ranking of "high" to this decision.

When assigning the difficulty scale, the team relies on its knowledge of the organization's current practices, experience with making small or major organizational changes, and the range of potential ways to best inform a decision maker. The identified activities are just a first draft of potential activities to include and may change when the range of other factors are considered when finalizing the plan, as will be discussed in the next chapter.

Depending on how many gaps and the amount of information available, the two teams in these examples may alter the ranking scales as they see fit. High, medium, low, or some other gradation may be appropriate. Another approach may be to rank order the risks and difficulties from highest to lowest.

7.6.3 Summary of decision ranking for the Enthusiast Series example

Table 7.2 gathers the Enthusiast Series example decisions, mentioned in Section 7.4, adding their relative risk posed to the program and the relative difficulty of informing the decision maker(s).

Keep in mind that the actual list of inputs for the plan development will be much longer. In practice, this extensive list reflects the range of necessary decisions to create a reliable product.

Source	Note	Decision(s)	RL	DL
1 Mandate	The Director of Engineering is requiring electrical engineering work to include adherence to a new derating guideline.	Has the team adhered to the new derating guideline?	H	L
2 Mandate	A key customer is requiring the use of a specific coating and finish that the organization does not have experience using.	What challenges will the new coating and finish pose related to reliability? Have we met the customer requirement for the specific coating and finish?	L	L
3 Regulatory	A new requirement for bicycles sold is to have a nonremovable serial number or embedded identification device.	What reliability challenges will the nonremovable serial number present that require a solution?	H	H
4 Gap	Conditions of use profiles are used to drive design and testing; however, they do not represent the most severe conditions that are anticipated.	Engineering (transaction-level) decisions are related to the design margin and safety factors based on expected use profiles and environmental conditions.	L	H
5 Gap	Current supplier selection is based mostly on cost parameters, with insufficient consideration of supplier reliability and performance capability.	Strategic, operational, and transaction-level decisions made on which vendors to engage for the project.	H	L

Source	Note	Decision(s)	RL	DL
6 Gap	The most prevalent field failure mode for Enthusiast Series bicycles is the chain slipping off the derailleur, and there is no fix for this problem. There is no current effort to make the derailleur subsystem more robust in future designs.	What are the root cause analysis results and potential solutions to the chain slipping off the derailleur?	H	H
7 Gap	A review of design guidelines for Enthusiast Series design engineers reveals no mention of tasks or activities that improve reliability.	How should the design guidelines for the design engineers be changed to include expected reliability improvement activities?	L	L
8 Gap	Enthusiast Series customers will be expected to maintain their bicycles at regular intervals, but the existing published preventive maintenance procedures are inadequate.	How should the published preventative maintenance procedures be changed to become an adequate resource for Enthusiast Series customers	H	L
9 Strength	There is an approved list of vendors based on criteria, including field data analysis, performance capability, and audits.	Is reliability performance represented well in the approved vendor list?	—	—
10 Strength	Plans are in place to train the Enthusiast Series team on reliability statistical methods.	What training is needed to develop in-house expertise in reliability statistical methods?	—	—

Source	Note	Decision(s)	RL	DL
11 Product life-cycle guideline	The product life-cycle guideline requires all product requirements be met (including reliability objectives) before the start of production.	Does the design meet the reliability objectives?	H	L
12 Requests	The mechanical engineer developing the frame is considering a new alloy and is not sure of the appropriate welding technique to ensure that the frame structure has the ability to meet the reliability objectives.	Which welding technique should be deployed in manufacturing the bicycle frame using a new alloy?	L	H
13 Interviews	The procurement manager mentioned an upcoming change of vendor.	What is the reliability risk associated with changing tire suppliers for this project?	L	H
14 Interviews	The engineer designing the new shipping containers for the project says the goal is to reduce shipping-induced damage and would like a breakdown of past out-of-box failure reports.	What are the prior salient failures resulting from shipping-induced damage?	L	L
15 Team discussion and meetings	The procurement manager mentions a change in sprocket lead time.	What changes to the prototype testing schedule will we have to make given the four-week increase in sprockets lead time?	H	L

Table 7.2. Decisions and their importance for the Enthusiast Series example. Key: RL Risk Level, DL Difficulty Level, H High, L Low, and "—" stands for Not Applicable.

7.7 Visualizing and prioritizing the decisions

One way to organize the gaps and associated decisions is to plot them on a 2 × 2 matrix or another suitable matrix. Label the horizontal axis "Risk Posed to Project," ranging from low risk to high risk, and label the vertical axis "Difficulty to Address" with an easy to hard scale. Not all high-risk decisions have the same magnitude of risk, so a good practice is to rank order these decisions from lowest to highest risk. Repeat the exercise for the low-risk decisions. Do the same for the difficulty levels as well. Next plot the high-risk and low-difficulty decisions in relative location to their rank within the two scales within quadrant 1. The highest rank-ordered decision for risk would be plotted farthest to the right within the first quadrant. Likewise, the least difficult ranked decisions would be plotted lowest within the quadrant. Repeat this plotting process for each quadrant. Plot the top 50% of the risk-posed gaps to the right of the centerline along the risk-posed scale and the top 50% of the most difficult to close above

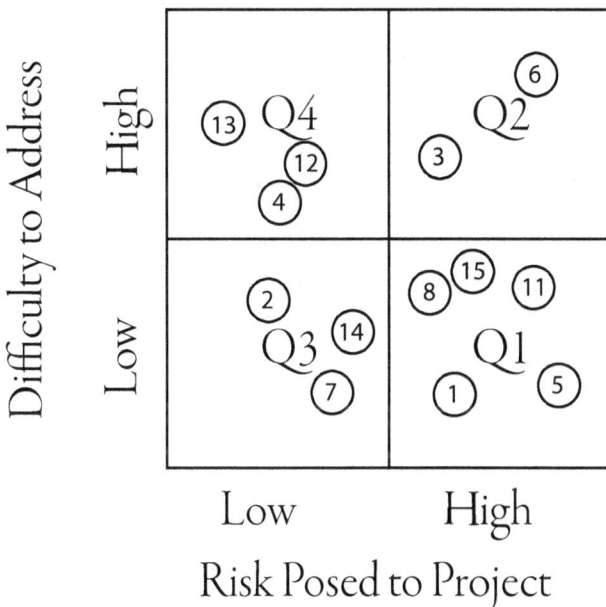

Figure 7.1. Decision-making matrix for the Enthusiast Series example.

the difficulty centerline. Figure 7.1 is the result, after the Enthusiast Series teams rank-ordered risks and difficulties.

The plotted numbers correspond to the numbered list of decisions (Table 7.2). If the two scales are rank-ordered, you can position each point relative to others within the same quadrant. For example, decisions with a relatively higher risk to the project would be plotted to the right of lower risk decisions.

The plotted decisions naturally fall into four quadrants. The lower right, quadrant 1, includes decisions that pose a high risk to the project and have a relatively low difficulty, meaning that they require a low investment but offer a high return. Quadrant 2 in the upper right contains the decisions that also pose a high risk to the project and may be difficult to inform adequately; these have high difficulty, high investment, and high return. Quadrant 3 in the lower left contains decisions that pose a low risk but are relatively easy to inform. We call these low-hanging fruit low-investment and low-return investments. Quadrant 4 in the upper left contains decisions that pose a low risk But are relatively difficult to inform. These are high-investment, low-return decisions.

To prioritize these decisions, consider the quadrants in order, 1 through 4. Using the criteria of prioritizing the activities to close the highest risk decisions and knowing that your resources are limited, we suggest starting with the decisions in quadrant 1. Decisions in quadrant 1 are high risk to the program and easy to address. It makes sense to prioritize these decisions first.

Next, consider the decisions in quadrant 2. Do any of these decisions risk preventing or overly burdening the team's ability to create a reliable product? Are any of these decisions showstoppers? That is, by not addressing the risk to the project, will the project be in jeopardy of failing to achieve reliability objectives? The showstoppers will require solving or, at a minimum, reducing the project risk. For decisions that are not showstoppers, consider what steps could be taken to start the work to partially inform the decision makers. It may take multiple project cycles to complete, yet each step may reduce the project's risk.

Quadrant 3 has the low-hanging-fruit decisions: These pose low risk but have low difficulty. Fit these into the prioritized list next. Consider which of these decisions, when informed, would provide strengths in the future that may make addressing the difficult and high-risk items easier in the future. These may also provide quick successes to create momentum to assist with more difficult challenges within the project.

Decisions in quadrant 4 are prioritized last. Consider which decisions may benefit from the activities that are likely to inform decisions in the other quadrants, providing some reduction in risk to these risks as a byproduct of creating the information (executing the plan tasks).

7.8 Summary

The reliability performance of an item is the cumulative result of the decisions made during the product's life cycle from concept through obsolescence. Focusing on influencing these decisions allows the development of a reliability plan that is useful, practical, and valuable. Remember that a reliability plan is not just a list of activities. It has to be connected to the needs of both the organization and the customer. Focusing on decisions is an important input to crafting a reliability plan.

In the next chapter, we will expand the discussion on the prioritized list of key decisions to include the other inputs necessary to select the vital few reliability methods that lie at the core of a reliability plan.

Chapter 8

SELECTING THE RIGHT RELIABILITY METHODS

I suppose it is tempting, if the only tool you have is a hammer, to treat everything as if it were a nail.

Abraham Maslow

In this chapter. Starting with the information we need to provide, the next step is selecting the reliability methods to include in the plan. A quick review of the decisions, constraints, and other factors leads to determining the most suitable methods that will provide the necessary results within the given constraints.

8.1 What is meant by the right reliability methods?

Every organization has a current set of skills and capabilities related to creating a reliable product. However, this set may not be sufficient to achieve the organization's and customer's objectives going forward.

The reliability plan will draw on current skills and capabilities, expand on current strengths, endeavor to close existing gaps, support key reliability-related decisions, and explore using methods not familiar to the team. This step in creating a reliability plan entails matching the most suitable reliability methods to best address each decision or requirement within the current set of conditions and capabilities. The idea is to select the best methods to provide the right information and create the most value during the process of creating a reliable product.

The question now becomes which of the many methods available to reliability engineers will best accomplish the reliability vision, close the gaps, inform key reliability-related decisions, while being respectful of the capability of the organization to improve.

8.2 Where does selecting reliability methods fit into the six-step process?

The focus of this chapter is step 4: Selecting the right reliability methods to achieve the reliability strategic vision given the current state of the reliability program with a focus on improving the outcomes of decisions that impact the resulting reliability performance of the product.

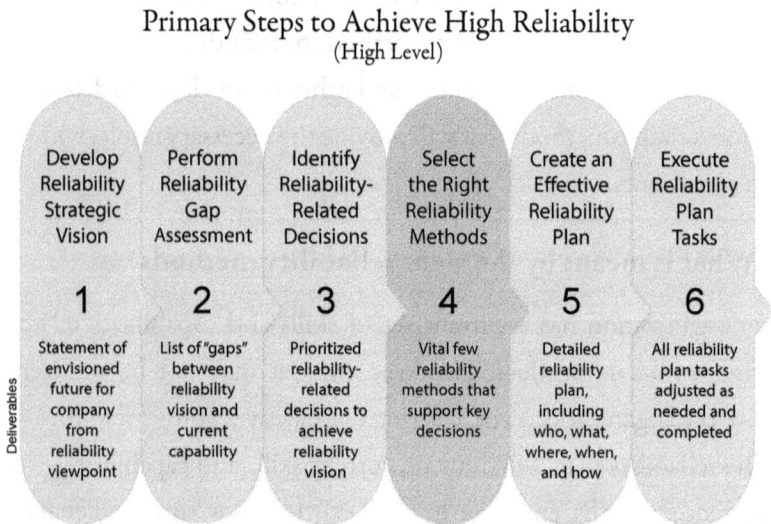

Primary Steps to Achieve High Reliability
(High Level)

Develop Reliability Strategic Vision	Perform Reliability Gap Assessment	Identify Reliability-Related Decisions	Select the Right Reliability Methods	Create an Effective Reliability Plan	Execute Reliability Plan Tasks
1	**2**	**3**	**4**	**5**	**6**
Statement of envisioned future for company from reliability viewpoint	List of "gaps" between reliability vision and current capability	Prioritized reliability-related decisions to achieve reliability vision	Vital few reliability methods that support key decisions	Detailed reliability plan, including who, what, where, when, and how	All reliability plan tasks adjusted as needed and completed

Deliverables

128

8.3 What are the inputs to reliability method selection?

To avoid doing what you always have done or using only the methods the team is familiar with, the inputs to selecting reliability tools and methods has to take a broader view. Considering the results of the first three steps is one set of considerations. In addition, the broader set of requirements, conditions, constraints, and opportunities facing the product development team must also be taken into account. The reliability plan has to integrate with the overall product development plan.

8.3.1 First three steps in the process to achieving high reliability

The reliability strategic vision, along with the specific project's set of reliability requirements and objectives, provides both short-term and long-term reliability objectives. It provides what the reliability plan should achieve when executed well. Remember, it's a good idea to "Begin with the end in mind."

The second step, the gap assessment, provides a set of current strengths and gaps that may support or hinder the team's ability to achieve the project reliability goal or the reliability vision.

The third step builds on this gap assessment to identify the key decisions that may benefit from supporting reliability-related information. The reliability performance of a product is the aggregate of the decisions made during the development process. Therefore, it is essential to provide the decision makers with timely results of selected reliability methods.

The prioritized list of key decisions, along with mandates and requirements from Chapter 7, provide a focus for the remaining steps. The list frames the specific objectives for the reliability plan.

8.3.2 Broader set of factors

The selection of reliability methods to include in the plan must also take into account a broader set of constraints and opportunities. These factors include safety, budget, timeline, assets, training, lessons learned, prior field reliability performance, and competitor reliability performance. This is a check step and builds on the results of the reliability gap assessment. While the assessment should include a broad set of factors, it is often focused on the current set of methods and the apparent gaps that limit the team's ability to achieve its reliability goals. This second look serves to prompt the due considerations of a range of pertinent factors that often have an impact on a plan.

The following factors are thought-starters to help discover any missing important inputs when defining the reliability plan.

Safety. Part of creating any product is considering and minimizing adverse or harmful consequences of assembling, transporting, storing, using, or retiring the product. This includes when working with a product during the design and development process. Consideration of material properties, potential consequences of failures, and safe methods to employ when bypassing a product's normal use protection and safety mechanisms all have to be addressed. Furthermore, the equipment and methods used to evaluate a system, prototype, subsystem, or element of a product must be considered, with the intent to create and maintain a safe working environment when assembling and evaluating a product or its elements.

Budget. It is rare to have sufficient funding for all desired tasks. The reliability method that could provide precise and complete information may require resources that are beyond the available budget or would preclude conducting other activities. Alternative methods may cost less and provide less precise or incomplete information, yet they may be sufficient. Being prepared with a basic cost–benefit analysis to support the plan's potential methods may provide a means to select the methods that provide the highest value for a given budget.

Timeline. Like a budget, it is rare for a product development team to have ample time to conduct all desired activities. Some reliability methods may prove better (e.g., more accurate) when conducted over an extended period; however, if the results are obtained months after the decision they serve to support has to be made, the longer method has little to no value. Consider the overall time available for the plan and consider when decision makers have to make the key decision such that the proposed reliability method would provide meaningful information on time for decision-making. Another consideration related to the timeline set of constraints is lead time; it may take time to obtain samples, testing resources, or other essential elements for specific reliability methods. Other methods may be available that may have shorter lead times.

Assets. Having a set of thermal chambers available is great if a reliability method supports a key decision requiring thermal chambers. However, selecting reliability methods just because you have the assets available misses the point of building a plan that provides value by providing support to decision makers. Knowing what is available both in-house and via third parties is still something to consider when selecting methods.

Training. Sometimes, the right reliability method for a specific situation may require someone with specific knowledge or skill for that method to be successfully implemented. Other times, those needed to employ a set of guidelines or methods do not have sufficient awareness or knowledge of the specific method. In both cases, consider what resources are necessary to obtain or execute the desired training.

Lessons learned. A common feature of most product development life cycles is a retrospective step to consider what went well and what needed improvement. Studying what was learned on previous projects may reveal additional opportunities to build on prior success or implement reliability methods to avoid prior mistakes.

Prior field reliability performance. Few products are truly 100% unique and thus have elements and processes that have been used with prior products. How did those products or elements of prior products perform? What worked well and what did not work as expected? Also, consider how

131

potential changes in use or environment may affect product performance. If the data are available, what were the weakest elements of prior products and why? It may be necessary to use prior product field data or failure analysis reports to address upcoming decisions.

Competitor's reliability performance. There are two aspects to consider: relative reliability performance in the market and the competitor's design approach that defines its reliability performance. If this information is already available, judge how it may properly inform decision makers. If not, and the missing information is essential to inform one or more key decisions, what methods would best create the information?

Stakeholder priorities. Who needs to approve and adopt the plan and what are their priorities? Crafting a plan to meet a budget when the key stakeholders are focused on the priority of the timeline will be a plan that misses the mark. Who needs to implement the plan's methods and what are their skill levels and motivations? Who needs to consider the plan's results and what are their priorities and preferred methods to receive the information? Whom do others in the organization look to for guidance and approval and what are their requirements to actively support the plan? Who takes credit for the plan's successes or suffers the fallout for plan failures? Who is going to champion the reliability plan as constraints change?

This is not a complete list of considerations. Your set of constraints and opportunities may differ. The idea is to consider program or project constraints when selecting reliability methods to help ensure successful implementation.

8.3.3 Potential reliability methods

A third set of considerations is the array of reliability methods that could provide the necessary information to inform decision makers as they address the identified key decisions, requirements, and mandates. For example, estimating a component's expected reliability performance could employ methods that range from a simple engineering judgment (a guess) to a detailed multiyear failure mechanism characterization study. It also could rely upon existing relevant field or vendor data or entail conducting an accelerated life test on the dominant failure mechanism(s).

Ongoing professional development is needed to improve knowledge of the myriad reliability methods that are available. Awareness is a first step. Being aware of a broad array of methods along with required conditions or inputs, in addition to cost, accuracy, and time and expertise requirements, first provides the team with more options to effectively execute methods that add value. Second, this awareness also helps to prevent using just the old trusty hammer to address every problem that comes along.

The following is a brief description of reliability methods categories that can be useful in selecting the most suitable reliability methods. Appendix C lists reliability methods organized by the categories (found online at accendoreliability.com/go/pre/). Although Appendix C is not a complete list of reliability methods, it can be useful as a thought-starter for potential

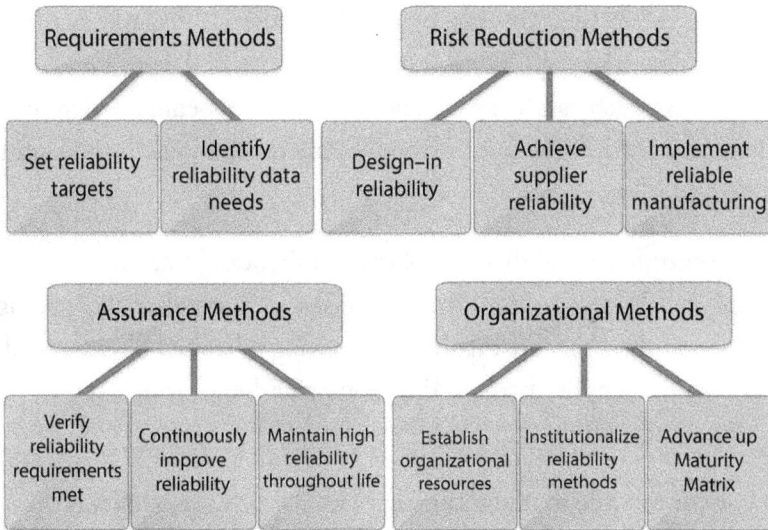

Figure 8.1. Reliability method categories and subcategories.

methods within each category. A summary of these categories is presented in Figure 8.1 and discussed below.

Requirements methods

There are two subcategories of requirements methods:

Set reliability targets. This subcategory embraces requirements-setting methods that need to be measurable at system, subsystem, and component level; verifiable during the product development timeframe; and converted into actionable technical specifications.

Identify reliability data needs. Reliability data are the fuel that supports measuring and monitoring progress toward attaining reliability requirements. Data can take the form of test failures, field failures, degradation measurements, test or analysis successes, and other forms. The focus needs to be on data integrity and correct measurements.

Risk-reduction methods

There are three subcategories of risk-reduction methods:

Design-in reliability. This category includes the DFR methods that can be executed in the design and manufacturing stages of the product development process, such as FMEA, physics-of-failure modeling, design margin analysis, and HALT. The focus of the DFR methods should be on high-risk and new design concepts.

Achieve supplier reliability. Supplier reliability can be achieved by incorporating reliability specifications and tasks into supplier bid packages, selecting suppliers that are capable of achieving reliability objectives, identifying critical supplier parts, and reviewing and approving supplier tasks for critical parts before shipment.

Implement reliable manufacturing. Manufacturing reliability is essential to ensuring that manufacturing and assembly operations do not significantly reduce the inherent design reliability of products. Steps must be taken to control manufacturing processes so that they are both stable and capable.

Assurance methods

There are three subcategories of assurance methods:

Verify reliability requirements. This category involves using physical testing methods and analytical modeling techniques. The focus needs to be on analysis and ALT. Supplier reliability requirements for critical parts should be verified before shipment.

Continuously improve reliability. The value of physical testing should be enhanced through the use of appropriate reliability growth models and life data analysis. A failure review system (FRACAS) can be instituted. Reliability improvement methods throughout the product life cycle must be maintained.

Maintain high reliability throughout life. Establishing and implementing proper service and maintenance procedures will extend product life and help ensure safe and trouble-free usage. Understanding and addressing customer issues during field usage is vital.

Organizational methods

There are three subcategories of organizational methods:

Establishing organizational resources. Organizational resources include personnel, training, procedures, and business processes. The organizational resource tasks will help move the organization to higher stages in the maturity matrix.

Institutionalize reliability methods. It is the responsibility of engineering to achieve reliability objectives. Reliability methods must be integrated into ongoing engineering procedures and tasks, including design reviews, supplier selection, work instructions, and design procedures.

Advance up the maturity matrix. It is important for any organization to improve its capability to achieve reliability objectives consistently. The maturity matrix represents advancing stages of reliability maturity.

8.3.4 Current capability

The current set of capabilities, as determined during the reliability gap assessment and summarized with the maturity matrix, provides another critical consideration when selecting reliability methods for the plan. Although there may be a "perfect" method for any given situation, implementing it may require expertise not available within the organization or equipment not likely to be available in a timely manner, given current budget limitations.

Selecting methods that build on the current set of strengths minimizes the investment of resources to create capabilities from scratch and helps bolster previous efforts to make improvements. Making incremental enhancements to capabilities leverages current capabilities and minimizes the resources necessary to make some improvement. Reviewing the maturity matrix may help identify small changes and/or recommendations that assist in improving the organization's reliability maturity.

The organization's current resource capabilities do not limit the reliability methods you may employ. They are just a consideration. Not having a temperature and humidity chamber does not mean you cannot use a rental unit or work with a local lab that has an available chamber. Moreover, just because you have a vibration table does not mean you have to use a method that needs a vibration table. You should select methods that are informed by capabilities—not methods that only use existing capabilities.

Factors to consider include

- expanding current strengths to more parts of an organization,
- renting capital equipment,
- hiring consultants with the necessary expertise (for the task at hand and for training), and

- investing in incremental improvements that enable improving reliability maturity.

8.3.5 Timing of method selection

Similar to the reliability gap assessment, the selection of reliability methods is input to a reliability plan. Therefore, it should be done early in the product development process, during and immediately following the identification of key decisions. It is important to keep in mind that the plan is not static, so this process should continue to gather inputs and drive you to select the appropriate methods as the project evolves and pivots, by learning and identifying other key decisions in need of reliability-related information.

8.4 What are the specific steps in selecting the most suitable reliability methods?

Selecting the most suitable reliability methods involves the following steps:

1. Gather the inputs.
2. Select potential methods to inform key decisions.
3. Evaluate and narrow the selection of draft methods
4. Review the list of draft methods with the team.
5. Document the relationship between selected methods and supported decision(s)

Modify the suggested process outlined here based on your experience and organization's business management processes. There are often many ways to generate information useful to inform decision makers, yet not all potential methods work well in all circumstances. Filtering potential methods with what is feasible and sufficient generally results in the identification of the right method for the current situation.

8.4.1 Gather the inputs

In addition to the array of considerations discussed in Section 8.3 above, gather details from decision makers that would define having sufficient information for their needs.

For each of the identified prioritized gaps and identified key decisions, include the nature of the reliability information expected or necessary to inform those making the decision. The information may be in the form of a statistical analysis, a comparison, a summary of observations, a survey, etc. Identify the criteria the resulting information should meet to be useful. This includes aspects of when the criterion should be met, how precise or certain it must be, in what format it is in, etc.

List key decisions and mandates with the expected or necessary criteria for the resulting information. Confirm with those making the decisions what they need to be properly informed when making the identified decision.

When gathering inputs, remember that it is helpful to prioritize the gaps, using criteria such as capability of implementation, early "wins" to build momentum, and the ability to enable informed decisions.

8.4.2 Select potential methods to inform key decisions

The intent of this step is to identify many reliability methods that may fully inform decision makers for each of the identified key decisions. This is an exploration of what is possible, with a focus only on those methods that could generate the necessary information. Brainstorm and list potential methods that may provide the necessary information. The idea here is to consider a wide range of possible methods that might work. Narrow the brainstormed list to those that do have the ability to create the necessary information.

Understanding the type of decision further narrows down the potential options for the correct reliability method. Some methods are well suited to provide the relevant information for specific types of decisions. Here is a summary of six different decision types:

1. Prevention: What can we do now to avoid failures or improve reliability?

2. Comparison: Which design, vendor, or procedure option is better given the known reliability?

3. Priority: Where should we focus our resources to best improve reliability?

4. Resources: Who and when should accomplish a specific task?

5. Objective: How do we set or what are the reliability and availability performance objectives, goals, or requirements?

6. Measurement: What is the reliability performance now or expected to be in the future?

See Appendix B, online at accendoreliability.com/decsion-types, for further details about each of these decision types. Appendix C, online at accendoreliability.com/reliability-methods, lists many reliability methods and notes the type of decision the method commonly addresses.

8.4.3 Evaluate and narrow the selection of draft methods

Narrow the list to those that also meet the additional criteria of timeliness, cost, precision, confidence, and capabilities (unless intended to build new or expanded capabilities) and will maintain or improve reliability maturity, that is, move to the right on the matrix. Adjust and combine methods when one method can provide information to two or more key decisions. Look for opportunities to save time and resources by using methods that, with some modifications, can inform two or more key decisions.

Create a draft timeline and check dependencies between methods to ensure that the inputs for each method are available when needed. This is also a check that all necessary methods are included in the plan as there may be a circumstance in which a method would benefit from the output of another method. Finally, select a set of proposed methods that have a high degree of certainty to adequately inform the key decisions and meet the range of other considerations and factors identified.

8.4.4. Review the list of draft methods with the team

A draft reliability plan, even just a listing of methods, benefits with a review by stakeholders of the plan. Invite decision makers, the project manager, lab managers, technicians, and those managing the eventual plans budget and execution to review the draft. Solicit input, including alternative methods to employ, whether anything is missing, whether any revisiting of assumptions is necessary, and whether there are any new or altered reliability-related decisions or considerations.

8.4.5. Document the relationship between selected methods and supported decision(s)

Record how each key decision will have its reliability information created and by which specific method(s). There should be no doubt which decision makers will receive which set of information generated by each method.

Every reliability method selected for inclusion in your reliability plan has a cost for implementation. Therefore, each selected method must be justified based on the ROI. Before including a reliability method, ask questions such as the following: What is the reason for adding the specific method in the reliability plan? What gap does it close? What organizational capability does it advance? What decision does it support? Does the method sufficiently identify or mitigate risk to justify the cost of implementation? Does the method help answer what is going to fail and when?

8.5 Lessons learned and additional considerations

Safety comes first! When it comes to reliability, always treat safety objectives as having the highest priority.

Creating an effective reliability plan gets easier with experience. Here are a few considerations that we have learned while creating plans in a wide range of situations. Selecting the right reliability methods is an essential step that sets the team on the path to achieving reliability objectives.

8.5.1 Be specific on method selection

If FMEA is a selected method, include how FMEA will be used and in what specific areas. If FMEA capability is a gap, include how the capability of performing FMEAs will be improved.

If DOE is a selected method, include how DOE will be used and in what specific areas. If DOE capability is a gap, include how the capability of performing DOEs will be improved.

8.5.2 Organize the methods to align with the four types of methods

This is a check step to help you avoid focusing on only one area to the exclusion of other facets of a robust reliability plan. Each reliability method typically works best with one of these four categories:

1. requirements methods,
2. risk-reduction methods,
3. assurance methods, and
4. organizational methods.

The first three align with the product life cycle from vision and goals establishment, through designing-in reliability, to product testing and continuous improvement. Organizational methods address business and information flow processes along with resources that support many projects such as failure analysis and product testing facilities.

Keep in mind that, while setting reliability goals provides the necessary information, if not checked or monitored, the goals alone will result in little influence on the design and development process. Also, it is often much less expensive to design in reliability early in the design process instead of discovering and solving problems later in the process—or worse, once the product is being sold. It is impossible to simply "test-in" reliability, considering reliability has to start early in the product life cycle.

Including proposed methods across these four categories is one more way to make certain that the plan is not missing a vital element

8.5.3 Additional considerations

Focus on the "vital few" reliability methods, thus avoiding including too many reliability methods in the reliability plan. The team can only accomplish a finite number of tasks.

Emphasize DFR tools as they assist the development of a reliable product right from the start. One way to justify the investment in addressing reliability early in the process is the "rule of ten" concept (Anderson 2020, p. 309), which means that changes made earlier cost less than the same changes made later in the product's life cycle.

As advised earlier, it is important to involve management in a meaningful way to support the development of the reliability plan, including the selection of reliability methods. Consider ongoing development of in-house reliability expertise to advise management on reliability methods and associated trade-offs. Foster improved expertise with training opportunities and encourage reliability professionals to seek certification. Working with management, subject-matter experts, and colleagues will require adept interpersonal skills. Gaining consensus on the right methods is necessary to successfully execute the plan and realize the value it may create.

Tailoring the reliability plan's selected methods to fit with the organization's maturity and capability is essential. The plan should build on what the team is able to accomplish, as well as assist the organization to learn and master necessary methods.

Finally, remember that each selected method in the plan has an explicit purpose and reason for being there. Each method selected will create results informing one or more decisions. Of course, as conditions change, the selected method may change or be removed from the plan as well.

8.5.4 Cautions

There is no perfect reliability plan. You will make mistakes creating plans, and that is a great way to learn. Here are few notes of caution that we have learned the hard way over the years:

- A common mistake is to use "canned" lists of reliability methods or copy the reliability plan from a previous project or program. Avoid the use of methods that do not have a good ROI.

- Challenge the use of methods to justify their inclusion in new reliability plans.

- You may encounter resistance if "favorite methods" are not included. Remain open-minded and flexible. There may be a sound reason for the "favorite method," or it may not be needed.

- Balance maximum value to improve reliability with the need to get the reliability plan approved and implemented. Choose your battles wisely.

8.6 Bicycle examples showing reliability method selection

We'll use the Enthusiast Series bicycle here to demonstrate examples of method selection. We started by randomly selecting three key decisions from the examples in Section 7.6.3 in Chapter 7, one each from mandates, gaps and strengths. For each of these key decisions, we will follow the steps for selecting the most suitable reliability methods.

8.6.1 Key decision 1 (from Chapter 7)

Table 8.1 summarizes a key decision to consider.

Source	Note	Decision(s)	RL	DL
1 Mandate	The Director of Engineering is requiring electrical engineering work to include adherence to a new derating guideline.	Has the team adhered to the new derating guideline?	H	L

Table 8.1. Key decision 1.

1. Gather the inputs:
   - The Director of Engineering mandate to adhere to the new derating guideline
   - A copy of the new derating guideline
   - "The Basics of Derating Electronic Components" (Schenkelberg 2022a) and "Creating meaningful derating graphics" (Schenkelberg 2017)
   - From the reliability strategic vision for the Enthusiast Series: a system reliability target of 95% at 7 years under the prescribed operating conditions

2. Select potential methods to inform key decisions:
   - Derating method XYZ that aligns with the new derating guideline
   - Derating analysis using MTBF
   - Supplier selection based on ability to meet reliability objectives

3. Evaluate and narrow the selection of draft methods.
   - Eliminate the derating analysis using MTBF, because of faulty assumptions about MTBF.
   - Supplier selection based on their ability to meet reliability objectives, although important, does not address the mandate from the Director of Engineering.

4. Review the list of draft methods with the team.

The derating method XYZ that aligns with the new derating guideline was reviewed with the project team. Feedback has indicated full support for derating method XYZ, but concern has been expressed about familiarity with the new set of guidelines.

5. Document the relationship between selected methods and supported decision(s).

Providing the development team with the XYZ derating guidelines enables it to apply derating of components in the design. Training or coaching on the new guideline should be provided as needed. Check on compliance during technical reviews, and report on compliance and any deviations from the derating guideline to the Director of Engineering after each technical review.

8.6.2 Key decision 2 (from Chapter 7)

Table 8.2 summarizes another key decision to consider.

Source	Note	Decision(s)	RL	DL
4 Gap	Conditions of use profiles are used to drive design and testing; however, they do not represent the most severe conditions that are anticipated.	Engineering (transaction-level) decisions are related to the design margin and safety factors based on expected use profiles and environmental conditions.	L	H

Table 8.2. Key decision 2.

1. Gather the inputs:
   - From the Enthusiast Series reliability strategic vision: "The enthusiast bicycle must be at least 99% reliable at 10 years of life, while being operated according to the functions and conditions of use from document XYZ."
   - Conditions of use document #XYZ

- From the Enthusiast Series reliability gap assessment: "Conditions of use profiles are used to drive design and testing; however, they do not represent the most severe conditions that are anticipated."
- Handbook of Reliability Methods (Note: an organization's internal document or suitable external document/reference.)

2. Select potential methods to inform key decisions.
   - Develop environmental conditions of use profiles for all bicycle riders.
   - Develop environmental conditions of use profile for targeted Enthusiast Series riders
   - Perform a worst-case analysis.
   - Perform a design margin analysis.
   - Perform a root sum squared statistical analysis.

3. Evaluate and narrow the selection of draft methods.
   - The current conditions of use profiles do not represent the most severe conditions anticipated, and good reliability requirements need well-defined conditions of use operating profiles. Therefore, create an environmental and use set of profiles specifically targeted at Enthusiast Series riders. Gather data via experimentations, observations, and instrumentation to characterize the salient stresses the system experiences fully. Clearly define the most severe conditions along with the range and nominal conditions. Incorporate these into product specifications.
   - Because we need to understand the most severe conditions and their impact on the bicycle and rider, perform a worst-case analysis.
   - Based on the newly created conditions of use profiles, perform a design margin analysis.

4. Review the list of draft methods with the team.

Review the narrowed selection of methods with the project team. As feedback, they agreed to the plan for developing conditions of use profiles targeted at Enthusiast users, including the most severe conditions. They asked whether the worst-case analysis would result in expensive components that exceed the budget.

5. Document the relationship between selected methods and supported decision(s).

The selected methods will provide critical input to design margin analysis and safety analysis. Document the results of the study in the conditions of use document #XYZ and in the product specifications document. Inform the development team of the availability of updated profiles by the end of the concept phase. Review the impact on program timing.

8.6.3 Key decision 3 (from Chapter 7)

Table 8.3 summarizes a third decision to consider.

Source	Note	Decision(s)	RL	DL
10 Strength	Plans are in place to train the Enthusiast Series team on reliability statistical methods.	What training is needed to develop in-house expertise in reliability statistical methods?	—	—

Table 8.3. Key decision 3.

1. Gather the inputs.

From the Enthusiast Series reliability gap assessment, one of the strengths is the plan to train the team on reliability statistical methods.

2. Select potential methods to inform key decisions:
   ◆ Short course from a selected university on reliability statistical methods
   ◆ Training by an outside consultant on reliability statistical meth-

ods (with the exact training being described in the reliability plan)

• Self-study material on reliability statistical methods

3. Evaluate and narrow the selection of draft methods.

After evaluation, the best option is training by an outside consultant on reliability statistical methods.

4. Review the list of draft methods with the team.

The project team agreed to the training by an outside consultant.

5. Document the relationship between methods and decision(s).

Once topics and dates for the training are set, inform the development team and other potential students. Improving the ability of the development team to use statistical methods can improve decision-making based on the data and analysis. Check on the use of statistical methods and document where they are used and to what effect or value.

As noted above, this is only a partial example, using three key decisions from the examples in Chapter 7. If we were doing an actual reliability plan selection of tools and methods, we would use the prioritized set of decisions to select all of the needed tools and methods. Just for practice, referring to Section 8.5.2 above, can you identify the four categories into which the selected methods fall in this example?

8.7 Summary

Although it is tempting to use the reliability plan that created a reliability product during a previous development cycle, it is unlikely to work in the current situation. Learning to balance the options, capabilities, and constraints is necessary to select the methods that produce sufficient results to properly inform decision makers. The next step is to take the list of selected methods and add sufficient detail to create a detailed reliability plan, which is the subject of the next chapter.

Chapter 9

CREATING AN EFFECTIVE RELIABILITY PLAN

It seems essential, in relationships and all tasks that we concentrate only on what is most significant and important.

Soren Kierkegaard

In this chapter. We start this chapter with a working definition of an effective reliability plan. We then examine the steps to create one. After a review of the previous step's output, it is time to organize the plan, including the essential elements and details.

9.1 What is meant by an effective reliability plan?

A reliability plan is a document that defines the entire set of tasks that need to be accomplished on a project or program, including responsibility for execution, timing, and resources, to achieve the program reliability objectives. The goal of the reliability plan is to enable you to design and manufacture a safe and highly reliable product on time and cost-effectively.

The reliability plan can be written for a specific project, a series of projects, or an entire program. Plans at the project level tend to focus on achieving the project's objectives with some work to improve a program or organizational processes. Plans at the program level focus on improving the

149

infrastructure and reliability maturity in support of achieving long-range organization goals. In its essence, a reliability plan is a roadmap to move a project or company from where it is today to where it needs to be from a reliability standpoint.

The reliability plan typically begins with a summarized version of the reliability strategic vision, reliability gap assessment, and key decisions that need to be supported. At its core lies the set of technical and business tasks to achieve reliability objectives. Each task should include timing and responsibility, with enough detail to ensure that the task is well defined. The reliability plan specifically avoids a long list of tasks that may exceed the resources and capabilities of the company. A reliability plan is not merely

- a list of planned reliability tasks,
- a test plan for the product,
- a copy of a reliability plan template such as Mil-Std 785,
- a statement of intent to achieve reliability goals, or
- a bunch of general words about how to achieve reliability objectives.

The reliability plan takes time and effort to create. A good plan provides all involved with a clear set of directions that, when implemented, creates valuable results. The reliability plan can be a stand-alone document or integrated into the overall product development process. Either way, it must be fully supported and aggressively implemented.

9.2 Where does creating an effective reliability plan fit into the six-step process?

Creating an effective reliability plan is the fifth of the six steps taken to achieve high reliability. This step brings together the reliability strategic vision, reliability gap assessment, key reliability-related decisions, and selected reliability methods into a sound executable plan.

Primary Steps to Achieve High Reliability
(High Level)

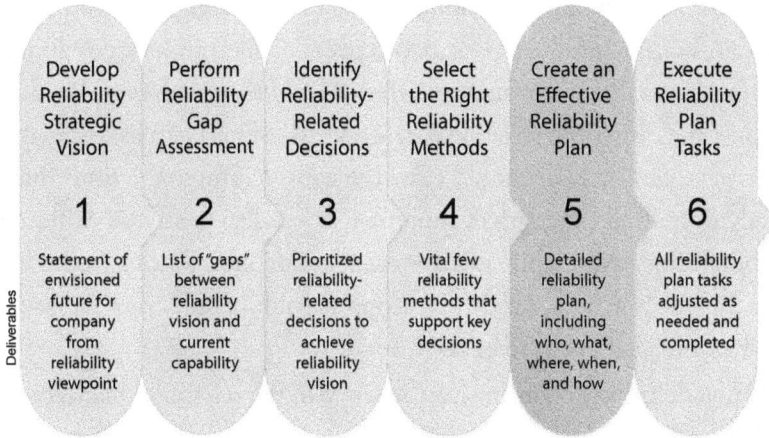

Develop Reliability Strategic Vision	Perform Reliability Gap Assessment	Identify Reliability-Related Decisions	Select the Right Reliability Methods	Create an Effective Reliability Plan	Execute Reliability Plan Tasks
1	**2**	**3**	**4**	**5**	**6**
Statement of envisioned future for company from reliability viewpoint	List of "gaps" between reliability vision and current capability	Prioritized reliability-related decisions to achieve reliability vision	Vital few reliability methods that support key decisions	Detailed reliability plan, including who, what, where, when, and how	All reliability plan tasks adjusted as needed and completed

(Deliverables)

9.3 What are the steps to develop a reliability plan?

Before writing the reliability plan, it is important to review steps 1 through 4 of the six-step process to be sure they are adequate, followed by defining or confirming the scope of the reliability plan. Once written, the reliability plan should be reviewed and approved by management. As with any plan, conditions and constraints will change over time. Moreover, additional information and risk will become apparent while executing the plan. Thus, the reliability plan requires regular review and adjustments.

Let's explore each step necessary to create a reliability plan.

Step 1. Review the reliability strategic vision, reliability gap assessment, reliability decisions, and selected reliability tools. This is an easy task if the earlier steps have been done well. The essence is to ensure that you have the required information available.

Step 2. Define or confirm the scope of the reliability plan. The scope of a reliability plan should be consistent with the business objectives and customer expectations. The reliability plan can support a single project, such as the bicycle Enthusiast Series for a single model year, or it can be much broader, such as all future bicycle projects. It can support all tasks from concept, design, development, manufacturing, field, and service, or it can be focused on one or two aspects of the product development cycle, such as DFR and assurance. The scope of the reliability plan should be consistent with the reliability strategic vision and gap assessment. An effective reliability plan should close the gaps, reinforce the tools that are already done well (strengths), support needed decisions, and achieve the reliability objectives, while incorporating guidelines, lessons learned, and ongoing best practices. Lastly, the scope should include the time horizon for when the plan needs to be completed.

When defining or confirming the scope of the reliability plan, consider the following questions:

- What are the boundaries?
- What is your charter?
- What are your constraints?
- Is this a project or program?
- Which aspects of the product development program need to be included?
- How do you know when you have a defined scope?

Step 3. Organize the essential elements of the reliability plan. Writing a plan starts with considering the needs of the audience. The plan provides confirmation of scope, requested resources for the supporting management team, and actionable details for the folks expected to execute the tasks. Reliability plans should begin with an outline of the content. The outline must be consistent with the defined scope of the reliability plan. This is an iterative process, building on what should be done to accomplish the vision and close gaps, yet bounded by what is possible given timelines and available resources.

The plan elements include a summary of the reliability vision, identification of major areas of concern, the rationale for task selection, and actionable task details. This outline provides a framework to fill in the specifics. The elements should be organized within a suitable format and style for your organization. The plan should fit within the constraints and be clear on the rationale behind the recommendations.

In the following, we give an example of outline topics for a reliability plan. This is merely one way to show the outline and is only offered as an example; it should not be used as a template. Any plan you devise needs to fit into the project management style of the organization.

To illustrate how merely using a template is not a good approach for a reliability plan, consider this story from Carl:

A company working for an original equipment manufacturer customer asked for my help with its reliability planning on a critical project. The first step was for me to go onsite, observe, and ask questions.

Question to management: "How was the reliability plan developed?" Answer: "Go down the hall and his office is on the right."

I walked down the hall and introduced myself to a reliability engineer. Question to the reliability engineer: "How was the reliability plan developed?" Answer: "It's a copy of Mil-Std 785. Here's a copy."

Question to the reliability engineer: "Did you get any support from engineering?" Answer: "No. The reliability plan is a requirement by the customer and filed with the program documents. There is no follow-up."

I marched into the management office and had a candid discussion explaining management's role in planning and achieving reliability. With the help of the project team, we performed a gap

153

analysis, developed a new reliability plan jointly with engineering, and ensured that the tasks were part of the job responsibilities of engineers and management.

As in most any situation, merely copying a template does not add value and cannot be used as a reliability plan.

Here is an example of an outline for a project-specific reliability plan (but do not consider this a one-size-fits-all template):

I. Summary
    Summarizes purpose and scope of the reliability plan.

II. Reliability Strategic Vision
    Outlines overall vision for reliability for the organization or program.

III. Summary of Primary Areas of Concern and Key Reliability-related Decisions
    Summarizes the reliability gap assessment, outlining gaps between the reliability strategic vision and current reliability capability and specifying the key reliability-related decisions that will be informed by the plan.

IV. Reliability Philosophy
    Provides a brief statement of the philosophy behind reliability task selection.

V. Reliability Plan Tasks
    Provides specific reliability tasks that must be done to close gaps and achieve the reliability strategic vision. Tasks must be well defined (who, what, when, how, etc.), and who will receive the resulting information from the task to inform which key decision should be specified.

VI. Resources and Training Plan

Outlines specific reliability plan tasks that support resources and training.

VII. Business Process Support

Outlines specific reliability plan tasks that support business processes and institutionalize reliability.

VIII. Reliability Methods Reference Section

Provides specific references for methods in the reliability plan as an appendix. The references may be internal guidelines or suitable external references to aid in the understanding and execution of the specific methods.

Step 4. Develop a set of tasks that will be included in the reliability plan. In step 4 of the six-step process, we have identified the vital few reliability methods that will be needed to close the gaps, reinforce what is done well, and achieve the reliability vision. In step 5, we transition from the list of reliability methods to the specific tasks, including a clear description of the tasks, who is responsible for implementation, when each task needs to be done, how each task will be accomplished and who gets the results, and how you know when each task is achieved (acceptance criteria).

Step 5. Write each task with sufficient detail. One of the keys to formulating effective reliability plans is writing the tasks at the right level of detail. The tasks should be executable at the "ground level," meaning by the engineers and staff who are designing, developing, manufacturing, and assuring the product or process, and detail the necessary resources.

Consider the following example. First, we should introduce the task with an explanation and provide context:

Develop and use a system reliability model for system ABC.

Definition: A system reliability model (SRM) is a graphical and mathematical representation of the reliability-wise relationship between the

155

subsystems and components of the system. The SRM is typically expressed as a reliability block diagram or a fault tree.

Application: Early in the product development process, the SRM will be used to flow down subsystem and component reliability requirements. It can then be used to make early predictions of system reliability. Later, during the reliability testing stages, the model will be updated as the basis for subsequent system reliability predictions and support reliability growth analysis. It supports two decisions, one related to identifying the lower level reliability requirements needed to achieve our overall reliability objectives and the other stipulating the status of reliability at each of the product development gates.

Next, we break down the task into executable subtasks, specifying the responsible person or department, the completion date for the subtask, and the resources needed. The actionable steps would look like this:

1. Ensure at least one person is trained in the proper procedure for System Reliability Analysis. Assign a person and provide training and software. (Responsibility, Date, Resource)
2. Determine the exact system configuration that will be used in the SRM. (Responsibility, Date, Resource)
3. Develop an SRM and use it to analyze system, subsystem, and component reliability requirements. (Responsibility, Date, Resource)
   a. Enter the current reliability requirements into the SRM at the system, subsystem, and critical component levels.
   b. Using the SRM, verify that reliability requirements flow down is feasible.
   c. Modify as needed to ensure the model reflects system reliability requirements.
4. Use SRM to analyze and predict system reliability as data becomes available. (Responsibility, Date, Resource)

a. When available, modify the SRM with failure mode frequency information.

b. As soon as test data are available, update the model with actual test results. The model can then be used to predict the impact of component test results on system reliability.

5. Continue updating the SRM on an ongoing basis, as a new test or as analysis data become available. (Responsibility, Date, Resource)

Add references, as needed, to provide sufficient detail for task completion.

Step 6. Review the plan and seek approval by management. As covered elsewhere, quality and reliability must be fully supported from the top of an organization. Part of this commitment is for management to guide, review, accept the specific tasks, approve, and allocate resources for the reliability plan.

9.4 Characteristics of well-written reliability tasks

Reliability plans should always be well written. The wrong way to write reliability plan tasks is to only state what methods are being implemented. Simply stating that an SRM needs to be created is not adding value. You have to stipulate the specific tasks needed to implement the SRM as done above. The characteristics of well-written reliability tasks include the details of each task as covered in the previous section. Just getting the tasks done does not necessarily add value on its own. Value is added by ensuring that the decision maker uses the task's results effectively. Remember to connect the method's output with the person who will be using the results. This is in line with the idea of adding value, ensuring the provided information makes a difference. See Schenkelberg (2014).

9.5 Bicycle example of a reliability plan

Using the Enthusiast Series example, the following are three of the selected tools from Chapter 8. We'll use these three examples to provide a portion of the reliability plan for the Enthusiast Series. Each task should specify the responsible entity, the completion date, and the resources needed.

9.5.1 Example reliability method 1: Derating method XYZ

Supporting decision: Has the team adhered to the new derating guideline mandated by the Director of Engineering?

Here are some example reliability plan actionable tasks:

1a. Clarify with the Director of Engineering whether there is a specific standard or guideline he or she had in mind for derating or whether one needs to be developed. (*Responsibility, Date, Resource*)

1b. If a new derating guideline needs to be developed, form a small team to develop the derating guideline. (*Responsibility, Date, Resource*)

1c. Gather and review published material on derating, including published standards and articles. (*Responsibility, Date, Resource*)

1d. Write a new derating guideline that meets objectives. (*Responsibility, Date, Resource*)

1e. Review the new derating guideline with the Director of Engineering. (*Responsibility, Date, Resource*)

1f. Once approved, roll out the new derating guideline with an education and reinforcement program including training, assessment, coaching, and reviews. (*Responsibility, Date, Resource*)

9.5.2 Example reliability method 2: Conditions of use profiles targeted for the enthusiast user

Supporting decision: Are engineering (transaction-level) decisions related to design margin and safety factors based on expected use profiles and environmental conditions?

Here are some example reliability plan actionable tasks:

2a. Identify the primary stresses and system experiences representing the customer usage profile. (*Responsibility, Date, Resource*)

2b. Gather data via experimentations, observations, and instrumentation to fully characterize the identified stresses. (*Responsibility, Date, Resource*)

2c. Clearly define 'the most severe' conditions along with their range and nominal conditions. (*Responsibility, Date, Resource*)

2d. Incorporate the usage profiles in technical specifications. (*Responsibility, Date, Resource*)

2e. Develop reliability requirements based on the well-defined conditions of use operating profiles. (*Responsibility, Date, Resource*)

9.5.3 Example reliability method 3: Training the team on reliability statistical methods using outside training consultants

Supporting decision: What training is needed to develop in-house expertise on reliability statistical methods?

Here are some example reliability plan actionable tasks:

3a. Make a list of potential outside consultants who can teach reliability statistics. (*Responsibility, Date, Resource*)

3b. Evaluate the merits of each training consultant on the list of candidates. (*Responsibility, Date, Resource*)

3c. Select the best candidate from the list of outside consultants. (*Responsibility, Date, Resource*)

3d. Schedule the reliability statistics training. (*Responsibility, Date, Resource*)

3e. Implement and evaluate reliability statistics training. (*Responsibility, Date, Resource*)

9.6 Summary

The cumulation of the work so far is to create a reliability plan. Each task in the plan has a clear connection to where its output will create value when used. No two plans are the same when created this way. However, each plan provides a way to generate the necessary information to create a reliable product and improve the organization's capabilities. The plan provides the road map, enabling the team to achieve the desired reliability goals and vision. The next step, the final of the six-step process, is to execute the plan, which is the subject of the next chapter.

Chapter 10

EXECUTING RELIABILITY PLAN TASKS

Our goals can only be reached through a vehicle of a plan, in which we must fervently believe, and upon which we must vigorously act. There is no other route to success.
Pablo Picasso

In this chapter. We emphasize the need to do what is in the plan to execute it. We then highlight several common obstacles you may encounter along with potential remedies. As with any plan, conditions change, so we have to be prepared to modify the plan accordingly. It is important to document what works or what doesn't and continue to improve your ability to create and execute an effective reliability plan.

10.1 The importance of executing the reliability plan tasks

Reliability plans have little value unless all of the tasks are fully executed. It is vital to follow up on each task to confirm completion to the satisfaction of the project team and to achieve reliability objectives and the reliability strategic vision. The reliability engineering team must bring problems with execution back to management.

The execution of tasks is not done blindly. As conditions and constraints change, tasks may become unnecessary, or tasks may need modification, or new tasks might need to be added. The overall objective of the reliability plan is to achieve the objectives, thus the tasks and priorities within the plan are likely to change.

While executing the correct tasks is essential, it is also important to execute tasks that are still important. Accomplishing the tasks provide the decision maker with the necessary information to achieve the reliability objectives, thus making the reliability plan valuable.

10.2 Where does executing effective reliability plan tasks fit into the six-step process?

We've reached the last step in the six-step process. Each step is important by itself, and the sixth step is no exception. Keep in mind the deliverable for this step: All reliability tasks are adjusted as needed and completed.

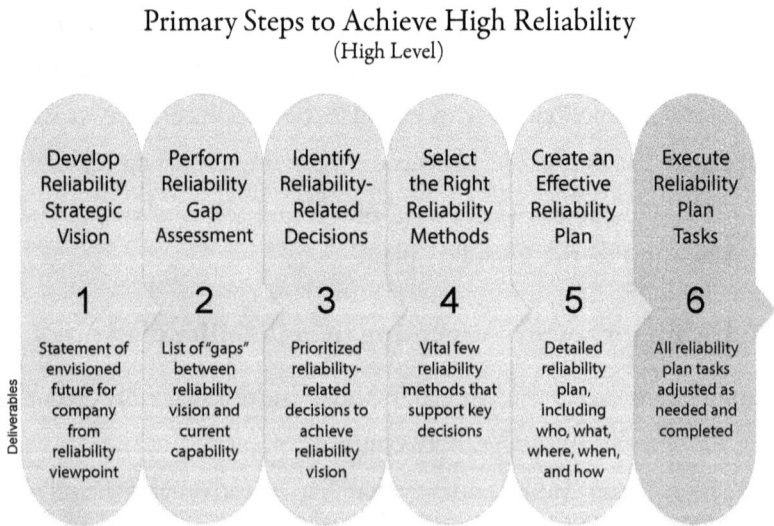

Primary Steps to Achieve High Reliability
(High Level)

	Develop Reliability Strategic Vision	Perform Reliability Gap Assessment	Identify Reliability-Related Decisions	Select the Right Reliability Methods	Create an Effective Reliability Plan	Execute Reliability Plan Tasks
	1	**2**	**3**	**4**	**5**	**6**
Deliverables	Statement of envisioned future for company from reliability viewpoint	List of "gaps" between reliability vision and current capability	Prioritized reliability-related decisions to achieve reliability vision	Vital few reliability methods that support key decisions	Detailed reliability plan, including who, what, where, when, and how	All reliability plan tasks adjusted as needed and completed

10.3 Project execution

If you have done a thorough job in creating an effective reliability plan, you have identified the set of tasks (including what, who, when, and why) that support the vital few reliability methods needed to accomplish the overall reliability objectives. Properly describing the details of the subtasks makes it easier to accomplish their execution.

In the previous chapter, where the reliability plan is created, management should have been involved in the process from the beginning and has approved each of the steps in the plan. Executing the reliability plan tasks requires the assignment of resources, which is why management needs to be involved from the beginning. The wrong way is to develop the reliability plan and bring it to management for execution. The right way is to have management involved from the beginning, supporting the reliability strategic vision, gap assessment, identifying key program decisions, and distilling these to converge on the vital few reliability methods. If management has been involved from the beginning, it is more likely to support the resources needed to execute the reliability plan tasks.

In most companies, there are processes in place to execute project tasks. Where possible, the best strategy is to leverage existing project execution systems. The reliability plan tasks should be integrated into the existing project management systems and supported by the project management team. If company project execution systems do not exist or are inadequate to support the timely execution of reliability plan tasks, the person or persons responsible for the reliability plan execution will need to work with management to gain support for the needed resources. This takes communication at the management level and persistence at the execution level.

10.4 Execution obstacles and remedies

Executing any plan of action takes persistence and tenacity. Reliability plan execution is no exception. Based on our experience, Table 10.1 presents some potential obstacles we have seen and how to remedy them. We'll associate the corresponding step(s) of the six-step process for clarity.

Obstacle	Remedy
Discover a new significant gap, e.g., a measurement system that is not capable (step 2).	Revisit the gap section and follow through to see whether it is important and modify the reliability plan accordingly.
Test measurement system fails (steps 2 and 6).	Evaluate the impact on data collection; redo or rework experiments to ensure good data and results. Evaluate the impact on the schedule.
Decision criteria changes (step 3).	Change in management roles, responsibilities, and priorities can alter corresponding project or program decisions. Because reliability supports decisions, the reliability focus or tasks may need to be modified accordingly. Maintain vigilance to ensure that each step in the reliability plan is still needed and correct throughout execution.
Schedule changes in the program, e.g., ship a month early (step 5).	Schedule changes are (unfortunately) a common occurrence. The key is to maintain focus on the objectives and not allow degradation of quality or reliability. The reliability plan must be flexible to potential timing changes, without compromising the end results.

Obstacle	Remedy
The reliability plan is written but does not get support from management because of competing priorities (step 6).	If the initial reliability plan is not fully agreed to by management, it is not likely to be supported in the execution phase. The remedy is to belatedly present the reliability plan to management and make needed adjustments until it is fully supported, including the necessary budget and resources. As discussed elsewhere, the best approach is for management to be involved from the beginning.
The reliability plan is written and has initial support from management, but the tasks are not getting executed, and target dates are missed, so the plan languishes (step 6).	If management agrees to the reliability plan, but the plan does not get executed, this is usually due to a lack of priority or lack of resources. The remedy is to bring the plan back to management and request higher priority and necessary resources. Execution requires persistence, and bringing the status of plan execution to management frequently is necessary.
Test samples are made incorrectly (step 6).	Evaluate whether incorrect samples are still usable, or create new samples and negotiate an updated timeline to deliver results.
There is change in focus or objective or there are budget cuts (step 6).	Management changes and wants a different focus or vision or cuts the budget. If this happens, stay in touch with key stakeholders. Adjust the reliability plan as needed and continue execution.

Table 10.1. Obstacles and remedies.

10.5 Modifying the reliability plan based on situational changes

Singer and businessman Jimmy Dean famously said, "I can't change the direction of the wind, but I can adjust my sails to always reach my destination."

Every day, changes occur that impact priorities, resources, timing, and other business and technical realities. These changes create both opportunities and challenges when executing the reliability plan. Reliability engineers and managers need to pay attention and take advantage of opportunities and make adjustments to the reliability plan tasks.

10.6 Verifying that tasks are complete and reliability objectives are met

There are two conditions that must be met for the reliability plan to be completed. The tasks must be fully executed, and the reliability objectives must be met. If the reliability plan is well conceived, based on the correct vision, and reflects a keen understanding of the gaps and strengths, accomplishing the tasks should result in improved decision-making, thus reaching the reliability goals and objectives. However, experience dictates that this is not always the case. Plans change as conditions, priorities, and information changes. An adjusted and well-executed plan that meets the agreed upon reliability objectives starts with the initial well-conceived plan.

10.7 Documenting wins and opportunities

During execution and following the completion of tasks, it is time to document lessons learned. These should include positives (wins) and negatives (opportunities for improvement). Identify what went well and should be reinforced in future reliability programs. Moreover, it is important to identify what did not go well and should be remedied in future programs.

Furthermore, document the plan's impact on key decisions related to changes that created value. If a plan objective is to reduce the field failure rate, monitor and document the actual change. If an objective is to reduce the chance of a program delay, identify the elements of the plan and how they contributed to reducing the risk of a delay. The intent is to build awareness of the value of making decisions that impact reliability performance that are well informed.

10.8 Summary

The execution step is where the plan becomes a reality. The plan will be adjusted as the set of constraints and information changes, which is normal. During execution of the plan, you and the entire team will learn the necessary information to enable better decisions. In addition, the organization learns to think about reliability methods and information in a different way, thus improving the organization's reliability maturity.

Although this is the final step in the six-step process, there is one other element to consider when working through the process: Each step requires working with other people. A product's reliability performance doesn't involve just one person, test, or report. Reliability is built into the product by decisions and actions by individuals across the organization. In the next two chapters, we will describe the set of critical skills necessary to influence individuals across the organization.

Chapter 11

BUILDING CREDIBILITY AND INFLUENCE SKILLS: FIRST STEPS

The wise man doesn't give the right answers, he poses the right questions.
Claude Levi-Strauss

In this chapter. We would be remiss by not including this and the next chapter on soft skills. Every step discussed in previous chapters includes working with other people. This chapter discusses the need for skills vital to build credibility and influence as you communicate and interact with others. This introduction is followed by descriptions, best practices or tips, and where to learn more of these five crucial skills: listening, questioning, speaking and presenting, technical writing, and learning.

11.1 The need for credibility and influence skills

There are many paths to becoming a reliability engineer. If you are adept with statistics, enjoy the detective work of failure analysis, or simply want to create a durable long-lasting product, you likely found yourself in a reliability engineering role. A science or engineering background is a great start. Time spent working with a design or maintenance team certainly helps. An advanced degree in reliability engineering is another path.

The element that is often missing, given an engineering background, is excellent soft skills. We know the engineering and science stuff: the formulas, the testing, and the data analysis. We can get stuff done in the lab or on the shop floor. Yet, to become an exceptional reliability engineer, or any type of engineer, the ability to communicate well is essential. Added to this is the ability to get your point across, enjoy credibility, and wield influence to help others understand and accept your proposals, ideas, and results.

11.1.1 Difference between hard and soft skills

We are constantly learning new skills, some of which we are naturally proficient at performing, whereas others do not come easy. The set of skills necessary to be a successful reliability engineer involves both hard and soft skills.

Hard skills include analyzing, understanding and working with the physical world. These skills include the science and engineering topics we learned in college, seminars, conferences, and on the job. Solving formulas, conducting experiments, and making data-based conclusions are examples of hard skills.

Soft skills, by contrast, involve our interactions with other people. They range from how well we interview to our ability to persuade others of a specific course of action. Just the facts and details based on hard skills alone are generally insufficient to get anything done. Soft skills include the idea of communicating and working well with others.

11.1.2 Deliberate practice

As with any skill, you can improve with deliberate practice. Deliberate practice is different from just practicing; it involves setting goals for the next step of improvement, getting meaningful feedback, and spending time learning from others. For example, you might rehearse the speech in front of a mirror out loud and alone when preparing for a presentation.

While this is a decent start, it would be better if you establish one or more specific goals for what you want to improve when making a presentation.

Let's say you want to improve your gesture ability to improve your audience's understanding. Invite someone you trust to watch you practice and actually give the speech. Solicit honest feedback on what worked and what needs improvement. If your coach is someone who speaks well while using gestures, that is even better.

For every skill in this chapter, there are online courses available for free or at low cost on edX.org, Udemy.com, Coursera.org, and Lynda.com (now part of Linkedin.com). A quick online search for communication or for one specific skill will also reveal many other options. Taking a course is not the same as deliberate practice yet it does provide you with information that can help increase your awareness of what you would like to improve.

11.1.3 Necessary soft skills

Credibility exists when others trust and believe you. Basically, it means you do what you say you're going to do and what you say is true and not deceitful or misleading. Influence is your ability to assist others in changing their mind, agree to a course of action, or consider another way to view the world.

For engineers, credibility and influence start with sound engineering skills. Then, building on that sound foundation, we add the abilities to

- listen well,
- question constructively,
- write clearly,
- speak effectively,
- discuss meaningfully,
- facilitate fairly, and
- learn continuously.

We will explore each of these and more in the following sections that may provide additional insights as you work to improve your skills.

11.2 Listening

Do you hear what others say? Do you really understand their meaning and intent? Or were you too busy preparing for your next verbal barrage? Consider the words of the Greek philosopher Epictetus: "We have two ears and one mouth so that we can listen twice as much as we speak."

One can learn a lot by listening. Really listening lets others know that you invite and value their input. Although listening may not seem to be a skill that one needs to learn or master, consider this: How often have you walked away from a meeting where one or more participants obviously were not listening? How often are points repeated in an effort to be heard?

Listening well, sometimes called active listening, is a skill that enables you to do the following:

- gather input, ideas, and information,
- let others be heard and understood,
- build trust,
- improve influence,
- encourage engagement, and
- discover ways to improve.

Being able to listen, and listen well, can be honed and improved. A focus on being a better listener will improve your ability to communicate and influence as a reliability engineer. It has benefits beyond our engineering work, too.

Every other soft skill builds on your ability to listen well. Listening allows you to be well informed as you learn, write, speak, and facilitate.

11.2.1 What defines a good listener?

A good listener is someone we enjoy talking with. A good listener is someone who actually hears and understands what others are saying or are trying to say. Active listening is more than just being physically aware of the sounds made by another person; rather, it is the mental processing and understanding of what is being communicated.

The art of communication involves the transfer of information from one person to another. If the receiving side is not listening, the transfer doesn't happen. When preparing to write or create a presentation, we often begin by understanding the audience. That involves listening.

To truly get the answer, we have to listen to the response to our question.

11.2.2 How to improve your listening skill

When in a discussion, you can take steps to improve your understanding of what others are saying and demonstrate that you are actually listening. Here are a few tips to consider and practice as you work to improve your active listening skills.

Establish the right mindset. Your frame of mind sets the stage for your ability to listen. If you are willing to learn, you are willing to listen.

Focus closely on the speaker. Devote your full attention to the person speaking. Avoid using your phone or computer screen, daydreaming, thinking about what to say next, etc.

Relax and breathe. If you are focused too intently, such as staring or leaning in, you may be inadvertently intimidating the speaker. Relax, smile, and take a breath. Focus closely, yet not too intensely.

Picture the words spoken. To avoid fixating on your memories or next set of statements or a story you wish to share, instead, as you listen, picture the words being spoken. This may be a literal visualization of the words, as if typed or written, or more conceptual in a mind map or outline.

Avoid interruptions and solutions. It is your turn to listen, so listen. Let the other person finish. Making assumptions and jumping to conclusions suggests you are not really listening.

Feel what the speaker is feeling. Are they sad, happy, fearful, or joyous? Then, be so yourself. Match the emotional state of the speaker as you listen. This will affect how well you understand what is being communicated.

Ask open-ended questions. Asking questions to expand the conversation and explore the other person's thought process and views encourages a continuance of the discussion. Avoid questions that require only a yes or no response.

Ask clarifying questions. If you do not understand a word or concept, seek an explanation.

Ask for details. Explore details by posing specific, direct questions on elements of the conversation. Ask "Tell me more about..." or "How would this work?"

Summarize and check understanding. When discussing a topic, you should occasionally summarize the points recently discussed and ascertain whether you understand the concepts expressed.

Encourage the discussion. Using body language (e.g., a smile or nod) or asking a positive question to show interest helps others be confident that you are interested and listening.

Grasp the total meaning. Conversations and topics can be complex. The words alone may say one thing, yet the body language may indicate something else. Look for congruency or the lack of congruency to fully understand the message. Irony and some humor may be difficult to fully understand.

Monitor your responses. Your body language, gestures, and comments may impede or encourage the discussion. Keeping an open mind and a willingness to learn and understand others with different views than your own may be a challenge.

11.2.3 Where to learn more about active listening

The book *Communication in the Real World* (2016) details specific steps one may take to improve listening competence, including a breakdown of the states of the listening process.

For an expansion of many of the tips above, see the article "10 Steps to Effective Listening" by Dianne Schilling (2012) in Forbes.

The technical paper by Levitt (2001) describes and defines active listening in detail. The paper describes the results of a set of observations relating the importance of active listening to the ability to perform other skills.

For detailed tips on improving your active listening skills, see the short online article "Active Listening" (Mind Tools Content Team 2022).

For a basic introduction to the concept of active listening, see the Wikipedia page "Active Listening" (Wikipedia 2022).

Carlson (2022a) wrote an article on the importance of active listening specific to performing the role of an FMEA facilitator: "Facilitation Skill # 5: Active Listening."

11.2.4 Summary

Communication is a two-way process. When someone else is talking in a conversation, you don't just wait for a pause to "jump in"; it's time to actually listen. You listen to understand the message being conveyed. You can learn to actively listen, thus improving your ability to understand, learn, and solve the real problems others share with you, not just the problems you "heard."

11.3 Questioning

Using questions is a skill employed to improve learning for the person asking the questions and to bring about knowledge transfer for the speaker or presenter. You can offer an answer as part of a presentation or pose a thoughtful question and ask the audience to consider the answer. Which way brings about better learning? Do you learn more by listening to a lecture or by interactive dialog? Or both?

One of the barriers to meaningful questioning is an exaggerated sense of self-importance. Plato talks about teachers using an "ignorant mindset" to compel students to explain things. It takes a bit of humility for the person who is speaking or presenting to step back and ask the audience or the listener a question to stimulate thinking before announcing the answer. Learning how to ask the right questions is a skill worth mastering.

11.3.1 What defines a good questioner?

A person who uses questions well does so to learn and to help others learn. Asking questions while actively listening may include clarifying or open-ended questions, both supporting fully understanding what the other person is communicating. Posing questions when learning helps you explore and understand new concepts, broadens your awareness, and invites additional information. By questioning your audience when teaching or presenting, you encourage engagement and dialog, thus facilitating an improved transfer of knowledge.

A person who asks questions well does so deliberately to improve the conversation, discussion, or learning opportunity. A great questioner poses genuine and honest questions and does not use questions to embarrass others or conceal information.

11.3.2 How to improve your questioning skills

As with any skill, the way to get better is to practice. Ask more questions. Specifically, ask more follow-up questions. Questions that elicit more information or deeper understanding help drive the conversation to a deeper level. Asking more questions does not mean just rapid-firing one question after another. As with any question, you should use active listening skills. After you have asked the question, you need to listen to the full and complete response. This often leads to the next appropriate question and continues the discussion.

Some questions may not have an answer. That is ok. Some questions may require additional information that may take some time to gather. Also, the person being asked the question may not know the answer.

Questions can be framed in many ways. Simple questions are used to merely seek clarification (e.g., What did you mean by the term "thingama-jig"?) Leading questions can invite further discussion (e.g., Would it have been helpful to conduct HALT earlier?). Open-ended questions can be helpful to facilitate continued engagement in a nonconfrontational way (e.g., What assumptions are being used in the system reliability model?).

As needed, ask clarifying questions. Ask if you don't understand a concept or use of a word. Asking for more information on a topic the other person knows you already know, or should know, may come across as deceitful or trying to challenge the other person. When you genuinely do not understand something, ask for clarification. A related aspect is to inquire about additional references so that you can learn more about the topic while not derailing the conversation.

Leading questions may help to move the conversion in a new direction, for better or for worse. Because they also can bias the answer to what is expected as a response, you should be careful about asking questions that evoke an expected answer.

Open-ended questions can help prevent the other person from feeling interrogated. It also allows the other person to provide unexpected information.

While you may prepare questions you would like to ask someone in an upcoming conversation, they should be asked casually as a natural part of the conversation. In short, don't read your questions, as this might come across as a test. Rapport matters to get clear and complete responses.

When asking questions in a group as opposed to during a one-to-one situation, you should be aware of the group dynamics. You may want to be cautious about or alter a response depending on who else is in the room. Sometimes people pose questions simply to enhance their status within the group. In such cases, you can reply politely, then redirect and focus your questions and listening on genuine topics in which the group is most interested.

Of course, offering up a flippant question in response to an honest inquiry is just rude. Asking a counter question is not a suitable technique to avoid answering a legitimate question.

11.3.3 We are all teachers

Good reliability engineers are also good teachers. Teaching skills can be practiced and learned and are enhanced by proper use of questioning. One of the core tenets of good teaching is to use questioning to augment learning. It is not enough to merely ask students, "Do you have any questions?" Ask students to put the principle being taught into their own words. Ask them how they could apply the principle. Be methodical in questioning each and every student in a balanced and respectful manner. People learn by thinking, doing, trying, and even failing. Enhancing your questioning skills will enhance your teaching skills.

11.3.4 Where to learn more about asking questions

A portion of the Coursera.org course titled "Improving Communication Skills" (Schweitzer 2022) specifically addresses asking questions, a deeper look at questions, and active listening. The entire course is very informative and highly recommended.

The Harvard Business Review article by Brooks and John (2018) examines how questions help you exchange information and improve interpersonal bonds.

On the site BetterUp.com, Perry (2022) discusses the elements of a good question, how to improve as a questioner, and how to get started asking better questions.

Harvard Business Review (2015) produced a short video that introduces the four types of questions you need when solving different types of problems.

Another short video on YouTube introduces five different styles of questions: open, closed, alternative, leading, and control questions (Expert Academy 2021).

11.3.5 Summary

Asking questions is a way to learn and build a stronger relationship with others. How you ask a question, given its context and intent, may convey interest in the topic or person and the honest desire to learn. Asking questions is also a way to transfer knowledge, which, after all, is one of our primary objectives.

11.4 Speaking and presenting

Whether engaged in a one-on-one discussion or presenting to a group, you need the ability to explain the problem and associated recommendations so that they are heard and understood. Presentation skills include how you use your voice, word choice, pacing, storytelling, persuasion, selling, and more. These skills enable you to facilitate learning, build unified teams, solve problems, and incite action. This may or may not mean using "slides" yet it does involve you speaking to one or more people. Presenting is much more than just having a slide deck.

Learning to speak well increases your confidence and that influences how you speak, as well as how others perceive you and the message you are conveying. Speaking well enables others to hear and understand your insights, observations, results, or recommendations. It improves your ability to engage and influence others to take action or accept your recommendations.

If your team nods off or drifts away as you present, or if your recommendations are rarely implemented, you likely need to improve your presentation skills. We often present proposals and reports. We talk about the plan or results. We want funding, approval, or action. We need to be proficient at making presentations.

The need for excellent communication skills is often one of the requirements for job openings. It is not there by chance. Your ability to communicate well, especially via presentations, is vital for your success and the success of your reliability program. If your team, management, or peers do not understand your proposal or report when you present, few will take the time to read the material instead. Your presentation skills provide the incentive for action from your audience, and this action can be guided using your presentation skills.

11.4.1 What defines a great presenter?

A good presenter is someone who prepares for the presentation, has a logical flow to the talk, and doesn't just read the slides or speak in a monotone. A great presenter does the basics, engages the audience, and motivates the audience to accept a recommendation. A great speaker has a clear message that is conveyed in a convincing manner. This may be done when responding to a question in a conversation or meeting, or it may be done using a prepared presentation on a stage.

The truly great speakers are confident. They know their material, the key message, the audience, and how to speak well. People actively listen to great speakers.

11.4.2 Understanding your audience

A big part of influence is understanding what is important to those you want to influence. If your proposed course of action impedes their ability to succeed, in their opinion, your proposed change is not going to be implemented. It is important to recognize that different people have different motivations. Some work for money, position, and power. Some work for the personal satisfaction of doing a thorough job, learning something new, or simply supporting some other endeavor.

With a development team, the motivations may be driven by cost, time to market, or technical wizardry.

The individual and team drivers may become tied together. For example, the team receives a bonus if we ship the product on time. Those motivated by money may work a bit harder to hit the shipping timeline. Those motivated by learning something new may chafe at the lack of time to explore new material.

11.4.3 Crafting your message

When making a recommendation, do your homework on the team's driving motivations and the key decision makers' motivations. Present the recommendations such that it reduces the risk of delaying a project, thus helping the individuals to receive their bonuses and help those so motivated to accept the proposal.

The challenging part is including how the proposal benefits those with different motivations as well. Knowing what is important to the team and to individuals helps you prepare and present the ideas in a way that is heard and acted upon by the team. For example, instead of recommending the team include two prototypes for HALT, we recommend we discover potential failure mechanisms earlier to have time to learn how to mitigate those failures. In this manner, we can minimize the chance of finding those issues late in the project, thus delaying the launch.

Connect the report, results, recommendations, and proposals to what is important to the team, business, customers, and decision makers.

Note that both understanding your audience and crafting your message apply equally well when writing.

11.4.4 How to improve your speaking skills

The key to improving your speaking skills is practice, practice, and more practice. Practice with audiences. Sign up for speaking engagements wherever you can and treat every presentation as another chance to learn. When possible, ask a friend to critique your presentation in a helpful manner. See the discussion on deliberate practice in Section 11.1.2.)

By far, the best piece of advice we can offer about speaking is to say less. Say what needs to be said—nothing more. If you have 20 minutes, do not try to cram two hours of material into the time allotted. Simply talking faster doesn't work. Again, say less.

Focus on the one key point you need to make. Think TED talks: These are limited to 18 minutes, and many are memorable, even those that take only a few minutes.

Tap into your personal interest and passion for the subject you are discussing. Nothing communicates better than personal passion. Lack of interest or passion can come across as boring and impede a connection with your audience, which is essential for knowledge transfer.

Part of the path to honing one's speaking and communication skills is to make a personal commitment that each person in your audience understands the material you are conveying. We've all attended lectures or been at meetings where the speaker shows up and delivers a message and walks away, not appearing interested in whether or not you get the points being made. A casual "any questions?" is not enough. Observing body language to ascertain whether the audience is grasping your message is very helpful.

Here is a short list of presentation elements that require some practice:
- Understand the audience.
- Understand the message or desired outcome.
- Understand the constraints around the presentation.
- Design the structure of your presentation (story, timeline, proposal, etc.).

- Create slides to support your presentation (not the other way around).
- Practice, practice, and practice (and get feedback, too).
- Breath, relax, and adjust the presentation based on audience feedback.
- Stay hydrated; pause to sip water as needed during a longer presentation.
- Practice and manage your time so that you are not rushed at the end.

Presentations are often accompanied by illustrative slides. Here are some tips to consider when preparing and using slides or any visual aids:

- The slide deck is not the presentation. It is not a report, a technical archive of your work, nor the entire contract, proposal, or plan.
- The slides or whatever you use as a visual aid is just an aid. They reinforce what you are saying. They may provide structure, graphical information, or humor, but they are always supporting the spoken portion of the presentation.
- Use a large font size, high contrast, very few words, and one concept or idea per slide. Remember that it needs to be readable from the back of the room.

How you speak matters. Here are a few tips to consider for improvement:

- Speak more slowly for a larger group.
- Vary your voice. Change pitch, speed, volume, etc.
- Vary your sentence length.
- Include pauses. A pause can convey vital information when done well.
- Breathe. Take a deep breath regularly during your presentation.
- Speak to individuals in the audience, and scan and look at your audience members.
- Do not read the slides or a script. You know the material—just talk about it.

- Speaking to individuals in the audience rather than reading slides brings about a connection, makes them feel involved, and aids in understanding your message.

Movement is also another aspect to consider. Here are some tips:

- Gesture in a natural and meaningful manner.
- Stand or sit tall with your arms at your side or bent 90 degrees (not in the fig-leaf position or crossed).
- If on a stage, move about if possible, again naturally and with meaning (but do not pace or wander aimlessly).
- Only use a pointer when absolutely necessary (e.g., do not point at every word or wave it around the slide or try to encircle something).
- If using a whiteboard or paper to illustrate a point, learn to draw and write clearly.

11.4.5 Where to learn more about speaking and presenting

Nancy Durante studied great speeches and presentations. She broke down the presentations into elements that can help us understand what made them great. Check out her books *Slide:ology* (2008), *Resonate* (2013), and *Illuminate* (2016).

Dr. Carmen Simon through the study of memory and brain science coupled with how to craft memorable presentations identifies 15 variables you can use to influence other people's memory. Her book *Impossible to Ignore: Creating Memorable Content to Influence Decisions* (2016) includes practical examples and proven methods to make your work impossible to ignore.

She also presents the concepts in her book in webinars, some of which are available online. They are worth watching for the content and how the presentations are crafted. Here are two recorded webinars, yet any of hers is worth watching: "The Neuroscience of Digital Content" (2022a) and "Deliver Memorable Virtual Presentations" (2022b).

Group Wizardry courses by Michael Grinder (2022) annually rotate through charismatic leadership, group dynamics, and presentation skills. All three are similar but have a different emphasis. All three include concepts such as working with or presenting to a group, nonverbal communication, and techniques such as pacing and matching.

Sharon Slayer wrote a book (Slayer 2011) after attending Group Wizardry and working closely with Michael Grinder. She focuses on how being prepared and aware allows you to improve every interaction you have with others.

Annette Simmons (2015) examines the power of storytelling and how to incorporate stories to improve your ability to communicate clearly.

Carmine Gallo (2016) also writes about the use of storytelling for effective and memorable presentations. She also examines what elements of storytelling separate the bland from the great.

Tony Jeary (2004) advocates that any time we talk with another with the intent to persuade them or even learn from them, individuals, or groups we are actually presenting. He then provides eight practical ways to improve how you interact with others.

11.4.6 Summary

There are numerous great resources on public speaking. Much of this advice you have been exposed to before. The key here is to practice. Get in front of audiences routinely and get feedback. Like many skills, you can improve with deliberate practice and active listening.

As outlined in this section, copious amounts of information, advice, and resources are available to help you become a great presenter. Doing so is important because, as reliability professionals, we often influence others through presentations at staff meetings, team meetings, and reviews. Our ability to present well improves our ability to be successful.

11.5 Technical writing

As a reliability engineer, you write proposals, plans, and reports. You write problem statements, report on failure analysis findings, and recommend process improvements. Your peers, teammates, and management want to understand your writing. They want to quickly get your point, find supporting information, and take action.

You write to document a process or plan. More often, you write to encourage others to take action. Writing clear, concise communications that incite action is a hallmark of a good reliability engineer. You are doing technical writing.

You can learn to write well, but, of course, this too requires practice.

The discussions above on speaking and presenting skills, understanding your audience, and crafting your message are equally important when writing.

11.5.1 What is good technical writing?

We write to convey information, persuade, inform, or educate. We write so that others have access to the information we are providing.

Technical writing includes our writing of emails, reports, proposals, white papers, reviews, and many other formats. You are performing technical writing when you write about a technical subject for others to read and comprehend. Good technical writing is concise, clear, and understandable. The reader should be able to understand the meaning or message, while quickly comprehending the logic and structure of the document.

We spend countless hours writing, from email to technical reports to reliability plans to proposals. Being clear is essential. Providing the main point or request early is crucial. Think executive summary. This increases the chance of someone reading the entire document. Making complete requests improves your chance of the request being implemented.

Simple things like writing in the active tense, editing, or basic spell-checking elevate your writing to become effective.

11.5.2 Best practices for excellent technical writing

Technical writing is not consumed as one reads a novel, from start to finish following the story. Some readers may only scan the document and read the summary. Others may dig deep into the supporting information and want to understand what is being proposed and the associated benefits and risks.

Use the various elements found in all basic writing. These include an executive summary (highlighting main points or results succinctly in the first section), a table of contents, informative section headings, well-crafted tables, and clearly labeled graphs; use appendices and indexes as appropriate.

A few other best practices to keep in mind include the following:
- Use the active voice where appropriate.
- Avoid jargon or cliches.
- Write conversationally.
- Write with your audience in mind.
- Include white space (avoid dense blocks of text).
- Include subheadings for easy skimming.
- Include descriptive captions on images.
- Edit, edit, and edit again. Ask someone else to edit your writing. (Running a spelling and grammar checker is not sufficient.)

11.5.3 Best practices for displaying data

Often technical writing includes graphics for the display of data or information. Just as with the writing, the graphics should be clear and concise. The primary elements (the elements that convey the meaning or message) of the data displayed should be uncluttered. Avoid decorations and enhancements that can obscure the intended meaning. For example, do not use a three-dimensional effect on a bar chart, as it does not add to clarity.

Other best practices include the following:

- ◆ Label the axes.
- ◆ Set scales to facilitate meaningful comparisons and understanding.
- ◆ Minimize the use of lines, rules, or borders to avoid interfering with the data.
- ◆ Refer to the graph in the text explaining the context and meaning.
- ◆ Use readable text within the graph, if text is necessary.

11.5.4 Where to learn more about technical writing

The best way to improve your writing is to write often. To really improve, get feedback on your writing (deliberate practice). Engage a friend or hire an editor to comment on your writing. The feedback will help you focus on where to improve.

There are numerous courses and seminars on writing. There are excellent online courses as well. Even local universities regularly offer writing courses often open to the public.

There are plenty of books on writing well, as well as guides. Don't forget to have and use the dictionary and a thesaurus.

Here's a short list of recommended books and courses:

Technical Writing Process by Kieran Morgan (2015)

The Insider's Guide to Technical Writing by Krista Van Laan (2012)

Words into Type by Marjorie Skillin and Robert Gay (1974)

Technical Communication, 13th ed. by Mike Markel and Stuart A. Selber (2020)

Write to the Point!: Letters, Memos, and Reports that Get Results by Rosemary T. Fruehling and Neild B. Oldham. (1988)

If writing a blog or article for online publication, see Pamela Wilson's books that cover in detail how to craft content that makes a difference:

Master Content Marketing: A Simple Strategy to Cure the Blank Page Blues and Attract a Profitable Audience (Wilson 2016)

Master Content Strategy: How to Maximize Your Reach and Boost Your Bottom Line Every Time You Hit Publish (Wilson 2018).

Edward R. Tufte's books provide examples and detailed guidance on how to present data or information elegantly and clearly:

Envisioning Information (Tufte 1990)

Visual and Statistical Thinking: Display of Evidence for Decision Making (Tufte 1997)

Visual Explanations: Images and Quantities, Evidence and Narrative (Tufte 1997)

The Visual Display of Quantitative Information, 2nd ed. (Tufte 2001).

Beautiful Evidence (Tufte 2006)

Seeing with Fresh Eyes: Meaning, Space, Data, Truth (Tufte 2020)

Any of Hans Rosling TED talks (2022) are wonderful examples of presenting complex datasets in an elegant manner. He created unique animated plotting styles and made the software publicly available. (vizabi 2022).

Getting the basics right when writing is important. While we often do not remember everything our 7th grade grammar teacher tried to teach us, there are references and guides available to jog your memory. Here is our go-to short list of such works:

Elements of Style by Strunk and White (1979)

Chicago Manual of Style (University of Chicago Press 2010)

The AP Style Guide, formally titled *The Associated Press Stylebook* (Froke et. al. 2020)

The Yahoo! Style Guide: The Ultimate Sourcebook for Writing, Editing, and Creating Content for the Digital World (Barr 2010)

11.5.5 Summary

We write for a reason. Our writing provides a means to convey information and should do so in a clear and concise manner. Beyond basic spelling and grammar, we can improve the comprehension of our writing by simply practicing. Write well and continue to deliberately learn and improve.

11.6 Learning

Learning may not resemble what many of us did in school. Learning is not about earning a grade or diploma. Learning is about understanding and internalizing knowledge. One learning model is about the learner moving from being unconsciously incompetent, to consciously incompetent, to consciously competent, and finally to unconsciously competent (Lane and Roberts 2022). This model suggests that to learn you have to start with recognizing what you do not know. Then you take steps to try, fail, and adjust (deliberate practice) until you have mastered the skill.

Learning may take many forms, from being curious and asking a question to working with a coach to master the ability to present well. Learning may start at being aware (consciously incompetent) of a topic, which allows you to prioritize which topics to pursue further.

11.6.1 What defines a good learner?

A good learner is someone who is never done learning. Good learners actively work to become aware of a broad range of topics, ideas, etc. They then decide on which topics to pursue additional learning. Finally, they can adopt and apply what they learn, and then learn from that to understand what they still need to learn.

Good learners explore a broad range of topics, including those inside and outside their field or background. There is no need to only read technical papers in mechanical engineering, as you might equally learn from a paper on software development methods that you can adopt and use.

Good learners use every opportunity to learn: in discussions with others; reading a newspaper, magazine, or blog post; listening to a podcast; or simply making observations as they move through their day. Learning to master a topic may benefit from deliberate practice. Finally, good learners have a thirst for learning and take time to improve their learning processes.

11.6.2 Best practices for learning

To master a topic—to have learned—is to create a meaningful change in your brain. In short, it means you have established a set of stable connections among your neurons within your brain. To do this may differ for different people and situations, yet there are a few tips (Shukla 2021) that may help you learn well.

Be deliberate. Have a purpose for learning. This may involve conducting research to understand an unusual failure mechanism, or it may be the desire to graduate, or it may be to satisfy your curiosity. The stronger the motivation and intent, the easier it becomes to learn.

Pay attention. Put down the phone, turn off social media, etc. Devote your full attention and your full concentration to the topic under consideration. This lets your brain prioritize and efficiently process the new information.

Consider the big picture and finer details. Consider the big picture of the topic and how it fits within the world. Likewise, consider the finer details and how these specifics relate to each other and to details in other known topics. The idea is to map the new information in different perspectives to improve recall and stability of learning.

Enjoy learning. Have you found the ease of learning a topic you enjoyed? Maybe the class or instructor set a tone of curiosity and liveliness. Having fun while learning is treated as a reward and assists the brain to process and retain information.

Consistently learn. Just as with continuous improvement, regular small steps in learning a topic makes a difference. This 'small-bite' approach enables your brain to mull over and fully catalog the new material. The old saying, "Learn something new every day" is related to this concept.

Take care of yourself. Another old saying involves the relationship between a healthy body and a healthy mind. The health and well-being of your social life, rest, fitness, and nutrition directly impact your overall health and well-being, including your ability to learn.

Try, try, and try again. Trial and error and using a variety of approaches create richer experiences and support mastering a topic effectively. The idea is to practice learning what works and what doesn't. Part of learning is learning what isn't right.

Examine the topic with others. Find others also learning the same topic to compare understanding. Find experts to discuss finer points and your understanding of the topic. How do others describe a topic? Here, books and videos may provide additional insights.

Build using metaphors and analogies. One way we learn is by using the scaffolding built by prior learning and experiences. It's easier to extend an existing scaffold than to build a new one. Therefore, when considering a topic, actively think about how the new material is similar to what you already know.

Fail sometimes. Part of learning is taking a risk, the risk you haven't mastered the topic yet. Learning also occurs when you are outside your comfort zone. Failure can happen. Learn from it and continue to learn.

Ask questions. Ask clarifying, probing, how-to, and other types of questions. Take an active role in learning by asking questions. Asking questions and finding meaningful answers will encourage connection between existing knowledge and the new information.

Teach. Preparing and discussing a topic so that others may learn is a great way to learn yourself. The feedback your students provide helps you understand which aspects of a topic you fully grasp and which require you to do additional work to improve your understanding.

Get feedback. Getting feedback is not the same as the grade on your paper in school. It is the notes and comments on the paper pointing out areas that are done well and those that are not so well. As discussed in Section 11.1.2, practicing with meaningful feedback helps you learn and master a topic. Having a coach or instructor is one approach when seriously trying to master a topic. Paying attention to informal or personal feedback is another approach, as is discussed in the next section.

11.6.3 Receiving feedback

Part of learning entails receiving advice, criticism, and other forms of feedback that may have the intent to help you improve. Consciously doing deliberate practice involves seeking regular and helpful feedback. Not all of us are fortunate enough to receive great feedback. We all do receive feedback, but some receive very little actionable feedback. A great way to learn and master a skill is to do deliberate practice, which involves receiving and accepting feedback.

If you offer proposals, give presentations, make requests, or even just ask for a favor, you will receive some form of response. It often is just an answer to the call to action and nothing more. At some point, you may be pulled aside by someone who wants to provide feedback on your behavior, delivery, ability, or skill. It is this type of feedback that is essential to your improvement.

You need feedback if you want to improve making proposals, giving presentations, or issuing requests. Asking for additional samples for a reliability test 10,000 times will not help you improve unless you receive actionable and meaningful feedback that helps you improve the next request.

Here are a few tips to help you receive, understand, and take steps to get better feedback and to improve yourself based on that feedback:

Lower your shields. The phrase "We need to talk" will instantly cause you to be wary, guarded, and defensive. Many other, more benign phrases also signal that the person wishes to give you some feedback.

When you realize you are about to receive what may be difficult to hear, the natural tendency for most of us is to take flight or fight. However, this response will prevent you from hearing and understanding the feedback being offered. It also makes others less likely to attempt to provide feedback in the future.

Recognize that offering constructive criticism is not easy for the person giving you feedback. Accepting the feedback will go a long way in making the other person feel at ease. Let the feedback provider tell you what he or she needs to say. Don't shoot the messenger. That someone has taken the time to offer feedback is a sign of the regard in which you are held.

Recognize that it's just information. Keep in mind that you cannot see yourself as others do. We all make observations about others. This unique perspective provides insights that you cannot otherwise obtain.

The feedback you receive is the result of an outsider's observations and assessment. It may or may not be extremely useful. Take the time to listen and understand, and check with others to determine the validity of the feedback.

Actively listen. Help the person offering feedback by paying attention, asking clarifying questions, nodding, etc. (See Section 11.2 on active listening for details.) Let the person know you want and value this candid feedback.

194

Get specific feedback. If someone mentions that your presentation yesterday wasn't very good, ask for specific examples that hampered the presentation. Was it word choice? Pacing? Not having a clear message? Or was it confusing? If so, how? Solicit examples, excerpts, or elements that help you understand what you can do to improve your next presentation.

Vague or general feedback is not particularly of value. You need to fully understand the specific aspects of the other person's assessment and feedback to act upon it.

Check your reception of the feedback. As with active listening, paraphrase your understanding of the feedback to check that you actually understand what the person intended you to understand.

Evaluate the feedback later. Suspend judgment. Like avoiding being defensive, this is difficult. When someone is providing you with feedback, it is not the time to judge the validity of the statements. The evaluation process may limit what you comprehend.

Focus on gathering the information and understanding the feedback. Take notes if you have to. Spend time later to process the feedback. Look at the information closely, yet later when you can focus on your interpretation of the feedback. Above all, do not challenge or dispute the feedback, or you risk shutting off future feedback.

Seek triggers that prompted the feedback. If the feedback was about a behavior that you would like to control, ask questions to help you ferret out the trigger that invoked the behavior. You may have a habit or typical response that is not having a suitable effect. To modify such a habit will require understanding the events or situations that invoke the undesired response.

Request constructive criticism. The next time you are heading into a meeting to request additional samples for a test, ask someone you trust to watch your presentation and request. Ask him or her to look for ways you could improve making the presentation and request.

Even after an event or presentation, ask attendees about their impressions and their suggestions for improvement. Ask for feedback. If you get a simple "It was good," ask for at least one thing that could be improved.

Learn from compliments, too. If someone remarks that you did a fine presentation, that is a compliment. Smile and say thanks, as will come naturally. Then enquire as to whether there is anything you could have done better. The above suggestions on listening, being specific, etc. all apply.

Decide on your next steps. Only you can decide what to do with the provided feedback. The options include ignoring the feedback, gathering more data, observing the behaviors in question, learning how to improve, and taking specific steps to actually improve.

Consider the following example, which is a story from Fred's experience:

> Years ago, about a month after starting work with a new group within a company, I spoke with a co-worker, who mentioned the lack of female pronouns during the interview. A short discussion ensued. After some thought about the feedback, I decided to deliberately use a balance of male and female pronouns in writing and speaking. To remember this objective, I set calendar reminders every couple of weeks. Then I asked our co-workers to pay attention to pronoun use during meetings, stories, and presentations.
>
> That was a memorable bit of feedback. A co-worker's thoughtful comment about routinely being aware of word choice and its impact on others sparked the need to learn and, with practice, continue to learn how to communicate effectively.

When you last received constructive criticism, what did you do? Did you really listen and attempt to understand the feedback? What works for you, or what do you wish you could improve?

11.6.4 Where to learn more about learning

A common refrain is to "learn something every day." Instead, work to expand your knowledge as much as you can every day. Besides learning how to listen, present, and write with every such interaction, you can also learn by being curious. Be prepared to learn, which implies you are willing to change your mind, enhance your skills, or discover a new topic.

Courses, workshops, seminars, conferences, books, videos, papers, etc. all provide a means to learn both from the content as well as from how the content is presented. Here are some recommended sources:

The three massive online open courses by Barbara Oakley and Dr. Terrence Sejnowski provide an informative and engaging insight into how people learn and how to be a better learner:

Learning How to Learn: Powerful Mental Tools to Help You Master Tough Subjects (Oakley and Sejnowski 2022a)

Mindshift: Break through Obstacles to Learning and Discover Your Hidden Potential (Oakley and Sejnowski 2022b)

Uncommon Sense Teaching: Teaching Online (Oakley and Sejnowski 2022c)

In their online course site, the folks at edX provide practical tips for successful online learning (edX team 2022).

A blog post by Aditya Shukla (2021) explains 16 different tips for learning based on brain science.

Peter Senge wrote about how organizations can learn in his book *The Fifth Discipline* (1990).

Learning is a creative process and learning to be more creative is a great way to be ready to learn. Mark Bryan, Julia Cameron, and Catherine Allen detail how to become more creative in their book *The Artist's Way at Work* (Bryan, Cameron, and Allen 1998).

Cameron Conaway (2022) wrote an insightful article in the Harvard Business Review on understanding and using feedback productively.

11.6.5 Summary

Receiving feedback can be difficult even in the best of circumstances. Not everyone is a sage coach who focuses only on your improvement. Be open to all feedback. This presents another opportunity to employ active listening skills. Being open to receiving feedback encourages others to provide meaningful feedback as they feel heard; they can notice that you (hopefully) work to apply the advice. You can both encourage and assist others provide feedback, thus helping others to help you.

11.7 Summary

Becoming aware that you can and should improve your ability to listen well, question effectively, speak and present confidently, write clearly, and learn always is just the first step in improving your ability to communicate well. The skills discussed in this and the next chapter enhance your technical skills as you work to craft and execute each reliability engineering activity.

Chapter 12

BUILDING CREDIBILITY AND INFLUENCE SKILLS: FURTHER STEPS

Leadership is influence. If people can increase their influence with others, they can lead more effectively.
John Maxwell

In this chapter. We continue to examine soft skills. These include making complete requests, discussions and conversations, meetings, facilitation, and change management. As in Chapter 11, we are using the basic structure of description, best practices or tips, and where to learn more. Combined with our technical abilities, mastering these soft and related skills enable you to effectively support and guide the creation of highly reliable products. Throughout this work there are references to the listed skills in Chapters 11 and 12. From listening well to facilitating a meeting, it is our work with others that allows the creation of a plan that, when executed, achieves the reliability objectives. How we work with others matters.

12.1 Making or receiving a complete request

Not every request we make is fulfilled. Not every assignment is accomplished. Not every task we assign is completed. Why is that? Possibly, because there is the lack of a complete request.

It may be that the person receiving the request was incapable of completing the task or decided to ignore it. More likely, it may be that our request was not clear. An unclear request increases the chance that the desired outcome will not be achieved. An unclear request is open to misunderstanding and confusion and can alter the path toward an unsatisfactory result. Understanding the essential elements of making a complete request improves the chance of achieving the desired outcome.

12.1.1 What defines a good and complete request?

There are seven elements involved in making and receiving a complete request: the requestor, the receiver, the action, the conditions or criteria, the deadline, common understanding, and competence. Let's examine them one at a time.

1. The requestor. Someone is making the request. If it's you, say so. If it is someone else, identify him or her. Be clear about who is making the request.

2. The receiver (requestee). Someone is the recipient of the request. If there is more than one person in the conversation, be clear about who is to receive the request.

3. The action. What is it that the requestor is asking the recipient to do? This is the request. It may be something as simple as "Please turn on the overhead lights," or it may be more complete, such as "Conduct a field data analysis on product line Tulip."

4. The conditions or criteria. The boundaries define what is an acceptable accomplishment of the request. If the overhead lights have a dimmer, you may need to define how bright to set the lighting. If the field data analysis is to include each model within the Tulip product line, say so. Likewise, if the analysis is to focus only on the last two years of data, say so. The conditions may include a format for the delivery of the results. For example, the analysis may require a report, a technical paper, or a presentation. Be clear concerning the scope, criteria, and expectations that make up the satisfactory accomplishment of the requested action.

5. The deadline. When should the person receiving the requested action accomplish the request? Set a specific date and time (e.g., noon on Friday January 22 or this Friday before the close of business).

6. Common understanding. If both the requestor and request receiver know the close of business on Friday is 5pm U.S. Central time, then they complete the details of what could be a vague deadline. For teams that work together, the time and date of the next staff meeting are obvious and known. If in doubt, be specific, or check the commonly accepted practices within the organization.

7. Competence. Is the person receiving the request capable of accomplishing the task? Do they have the necessary skill, talent, background, education, etc.? For field data analysis, does the person have access to the data, the appropriate data analysis software, and the ability to effectively accomplish the analysis?

12.1.2 The perils of skipping an element

In reviewing each element, consider what the omission of that element could cause to the eventual outcome. Not every request we make is fulfilled. Not every assignment is accomplished. Not every task we assign is completed. By omitting an element, we leave to chance that the person receiving the request will understand and accomplish the task.

Keep in mind that making a complete request enables the person receiving the request to ask questions, confirm or challenge assumptions, and otherwise form a complete opinion before accepting the request. Remember, we are making a request. We are asking someone to do something. That person is entitled to fully understand that request before accepting it.

12.1.3 How to improve your skill at making a complete request

Consider the requests, large and small ones, that you have made today. Were all the elements included, assumed, and understood? How was the outcome? Was the requested action done on time? Was it complete? Did it meet all the conditions?

What about each request that went well and helped all concerned to accomplish the requested action? For each request that did not go well, what element(s) was missing? You can learn from what worked as well as from what did not. Continue to examine and improve your ability to formulate complete requests.

Developing a facility at making complete requests takes deliberate practice, informed feedback, patience, and persistence.

12.1.4 Where to learn more about making a complete request

Here are a couple of handy resources on complete requests:

Fred expands on the elements of a complete request in "The 7 Essential Elements of a Complete Request" (Schenkelberg 2022b).

Along with many other ways to improve your ability to communicate well, the book *Smart Work: The Syntax Guide for Mutual Understanding in the Workplace* also examines requests and agreements (Marshall and Freedman 1995, pp. 119–128).

12.1.5 Summary

While a comment such as "Let's meet on Tuesday" has some elements of a request, it misses essential aspects that should be obvious. Recognizing all the elements of a complete request allows you to pose clarifying questions, check assumptions, and fully understand any requests you receive.

When asking someone or a group to act on a request, you should help them to fully understand the request to avoid misunderstanding and confusion. You can assist them in fulfilling the request by starting with a complete request.

12.2 Discussions and conversations

Our ability to engage in one-to-one and informal small-group discussions and conversations is one of the most common ways we gather information, explore ideas, prioritize problems to solve, and arrange where to have lunch next Thursday. Discussions may have a specific topic and a facilitator to guide the discussion. Conversations are often informal and without a stated agenda, in which case the discussion may explore one or many topics.

Discussions help to fully explore a topic, discover new connections or facets of a concept, or build consensus on a course of action. A discussion may have a set goal, whereas conversations often do not have a defined goal.

Some people naturally use conversations to form and explore ideas. Talking helps them think. Some people are thoughtful before speaking, while still others might attempt to dominate the discussion.

Having the soft skills to both understand and deal with the many discussion styles while still moving forward with the purpose of the discussion will set you apart from other hard-skill-heavy engineers.

12.2.1 What defines a good discussion or conversation?

In the following, we'll use the term "discussion" to mean both a discussion and a conversation for convenience, as most aspects of what makes a good one and how to improve it are very similar. A good discussion
*   is on point, two way, clear, honest, and respectful;
*   involves all parties engaging, participating, challenging, and laughing, with active learning and great questions; and
*   is an event where everyone benefits from the experience, being worth the time and enjoyable.

12.2.2 How to improve your discussion skills

The best way to get better at engaging in meaningful discussions is to practice. If possible, get feedback from someone you trust to be honest with you. One technique is to keep a journal or morning pages; for example, write three pages, longhand and stream-of consciousness, first thing in the morning every day (Bryan et al. 1998, p. 5). Reflect on the discussions held or coming up. What went well? What made you uncomfortable? What preparation do you need for upcoming discussions? What do you want to accomplish in your discussions? How can you best support or help others in your conversations?

Participating or leading a discussion or conversation with more than one other person adds the complexity of group dynamics. A group of people often exhibit a separate behavior that is in part a blend of the individual's behaviors yet is often very different from that of any one person's behavior. Topics, comments, and level of participation all vary depending on who is in the group and what are the understood group norms. Paying attention to how the group, as an entity, responds to a change in topic, an interruption, or the approach to consensus is as important as how individuals respond to such situations.

Talking is not the same as a discussion or conversation. Talking is unidi-rectional. If two people are talking, they are talking at each other, not with each other. A discussion, in contrast, is a two-way discourse. When two people have a discussion, information passes both ways. Both speak, and both listen.

We engineers have plenty to discuss. We work with others to find solu-tions, make compromises, determine optimizations, and finish projects. We need to share our knowledge and insights and learn from others.

You can learn to foster true discussions and minimize simply talking at one another. You can take steps to enable the give-and-take exchange of a discussion. If you practice listening to the conversations taking place around you, you will discover that many are not productive, useful, or meaningful. With a little practice, you can improve the chance that your next discussion will be productive, useful, and meaningful.

Here are ten elements to keep in mind as you prepare for and engage in your next great discussion:

1. Be a great listener. See Section 11.2 on listening and get that down first. You have to not only hear what others are saying but need to understand what they are saying. You can only respond meaningfully and constructively if you pay attention to what others are saying.

2. Be prepared. Know what is important to others. If you don't know, ask. If the conversation is going to be on a topic you know very little about, do a little research beforehand. Learn some terms or concepts. Being familiar with the topic, challenges, objectives, or interests of others in the discussion helps you engage deeper and ask better questions.

3. Build the relationship. This is likely not your first or last discussion with your peers, colleagues, or management team. For you, one aim of every discussion is for others to enjoy talking with you. If you are easy to talk with, others will want to talk with you. The topic at the moment may not be of direct interest or ability to move your goals forward, yet there will be other conversations. Being rude today will make engagement more difficult tomorrow.

4. Respect and value time. A useful practice is to ask, "Is this a good time to talk about...?" If the discussion should only last five minutes, do not drag it on any longer. If the person has to go, respect that and catch him or her next time. Moreover, if you have a meeting planned for an hour, and the necessary discussion only takes 15 minutes, end the meeting early—there is no law that states we have to fill the time scheduled.

5. Let them talk. People like to talk about themselves. Let them. Nod; ask relevant questions; if you are listening, you will learn a lot about them.

Sometimes they will help you solve problems, understand their position or constraints, etc. without the need for you to pose questions.

6. Ask how to provide support. When a conversation turns to challenges, hurdles, constraints, or similar topics, that is a great place to offer your support. Ask how you can help the other person move forward. Since you do have to actually provide the agreed-upon support or action, do not offer what you cannot provide. The trick here is not simply to offer but to deliver. This provides a sense of trust and obligation. Mostly, it improves the ability of future discussions to go deeper and tackle tougher issues.

7. Be the real you. Breathe, smile, and relax. You should not try to be someone you are not. Be yourself. Masks or false personas aren't of much use.

8. Remember key points. Remembering key concepts or ideas builds on listening. Refer back to what they said, not what you said. If you don't understand something, ask. Inquire as to whether you have the key points or takeaways right. This is best if worked into your conversation, as it shows you are listening. Avoid just repeating what they say or summarizing, and build on the topic in part by referencing the key points made by others.

9. Let your actions speak for your accomplishments. Instead of reciting your amazing resume or list of accomplishments, focus on what you can do to realize another great accomplishment with the members of this conversation. Instead of responding to someone mentioning the school they attended or the problem they solved with mentioning your school or the problem you solved, ask about his or her school or the problem, and carry the conversation forward. Avoid engaging in an "Oh, I once solved a bigger problem..." back-and-forth game. You have accomplishments, and, in part, great discussions facilitated those. Focus on the conversation at hand, not how great you were in the past.

10. Create a safe and welcoming opening for others. Start conversations with open-ended questions, interest in others, and suggestions for a topic to discuss. Be inclusive. If there are three or more people in the conversation, create openings for or acknowledgments to those not as active in the conversation. Pause. Invite others to join the conversation.

12.2.3 Where to learn more about discussion and conversation skills

Deutschendorf (2014) wrote an informative article for Fast Company that reinforces some of the advice presented here.

12.2.4 Summary

Improving your discussion and conversation skills does two important things. First, it enables you to learn from others. Second, it increases your ability to influence others. These skills build on active listening (as does most every skill discussed here), asking great questions, and being willing to learn. These combined skills still require doing your homework, loosely holding an agenda, and being willing and purposeful in both the topics you initiate and those in which you engage.

12.3 Meetings

In most companies, much of the day-to-day effort to move reliability projects forward takes place in meetings. Meetings can be effective forums to accomplish cross-functional work—or they can be a waste of time. In this section, we'll introduce the characteristics of well-run meetings and share a list of meeting norms. Readers are encouraged to do an additional study into the principles of effective meetings.

12.3.1 What defines a good meeting?

Some of the characteristics of well-run meetings (Bens 2000) include the following:

- Starting and ending meetings on time
- Publishing and sticking to agendas
- Developing and getting agreement on meeting norms

- Always maintaining focus on the meeting objectives
- Summarizing results and follow-up actions at the conclusion of the meeting
- Preparing required documents, visuals, network access, software, etc.
- Ensuring that decision-making options are clear
- Encouraging healthy member behaviors
- Providing periodic process checks
- Implementing a process to create true closure
- Providing detailed minutes and specific follow-up plans

12.3.2 How to improve your meeting skills

First, remember that a group is an entity of its own. How individuals behave is often different within a group and again different with different groups of people. The dynamics between individuals or between status levels influence individual behavior as they gauge their participation based on who else is in the room. Meeting norms for a group also provide boundaries or guidelines related to acceptable behavior.

Meeting norms are behaviors agreed upon by meeting participants. They need to be developed by the team or the company. Reliability teams can use predetermined templates and develop company-specific guidelines. Below is an example of what comprises a set of meeting norms.

It is expected that each meeting participant do the following:
- Arrives at meetings promptly as scheduled
- Respects others' opinions
- Debates differences of opinion calmly
- Takes responsibility for assigned actions
- Listens carefully to all ideas
- Avoids emailing or using cell phones or other personal devices during meeting time
- Maintains focus on the agenda
- Sticks to the agenda and does not go off topic

- Provides constructive feedback
- Maintains equal opportunity for participation by all team members
- Engages in no "war stories" or side conversations

12.3.3 Where to learn more about meeting skills

Although there are numerous books, courses, seminars, and workshops, with more becoming available every day, here are three articles that provide practical and insightful advice so that you can conduct or participate in effective meetings:

Rathore (2022) provides advice for running virtual meetings.

Rogelberg (2019) focuses on how to determine how bad your meeting management is and what to do about it.

Koshy (2017) et al. focuses on making meetings effective.

12.3.4 Summary

Meetings provide an opportunity to enhance effective communication. Whether facilitating or participating, we all have roles in enabling meetings to present a venue for sharing knowledge and solving problems. Meetings are another forum in which to practice active listening, questioning, and discussion skills. Using your skills, along with helping others improve theirs, enhances the productivity of every meeting.

12.4 Facilitation

Getting something done by committee, conducting a meeting to achieve the desired outcome, and managing a root cause analysis team are all examples of when you will need to exercise your facilitation skills. The soft skills required build on many of those mentioned above, while enhancing your ability to work with individuals and the group to achieve objectives.

Serving the group and the group's purpose will garner success more often as you work to facilitate a meeting or process.

The mix of hard and soft skills requires regular care and feeding. To stay sharp with engineering skills you need to do the research, attend seminars, and ponder engineering solutions to problems. Learning and mastering soft skills involves the same effort. You have to extort time and practice to recognize your current skill levels and identify which skill sets require improvement.

12.4.1 What is a facilitator?

A facilitator is "one who contributes structure and process to interactions so that groups are able to function effectively and make high-quality decisions. The facilitator is a helper and enabler whose goal is to support others as they achieve exceptional performance" (Bens 2000, p. 5). The facilitator's role is not a passive position but a proactive role, encompassing general leadership skills. It is important to know that facilitation and leadership skills can be learned.

12.4.2 Are facilitators neutral?

The question about whether or not facilitators can interject their opinions in meetings is a crucial one. If done without steering the team to one conclusion, there is no reason why facilitators cannot offer their own opinions. The team facilitator has a proactive role, encompassing general leadership skills. Unfortunately, most companies do not have the option of hiring professional facilitators, as teams are often small and lean. The best practice is for the team to be led by a trained and experienced facilitator who also brings his or her own subject-matter expertise to the meetings.

12.4.3 Basic facilitation guidelines

As reliability professionals, we often lead teams to identify and mitigate risks. We help cross-functional teams identify and implement solutions.

We bring people together and support their ability to communicate clearly with each other. We facilitate.

Whether a leader or participant, we have a role in achieving the desired goals. Our ability to facilitate enables us to work with others to get things done. Understanding how to facilitate well enables us to add value when leading or participating on a team. Preparing for the facilitation role entails both planning the meeting and preparing yourself for the tasks ahead. Understanding the objectives of the meeting or session allows you to craft an agenda along with specific activities that focus on meeting these objectives.

The role of a facilitator is to help the group to achieve its objectives easily and in a smooth manner. This is done by first setting aside any of your own objectives or opinions that may interfere with the group's discussion, discovery, and achievements. If you are unable to act as a neutral party on the topic, you need to step aside and find a suitable alternative facilitator.

The following are basic guidelines to which you should adhere:
- When facilitating, stay neutral and in service of the group and its objective.
- Guide the meeting and help all participants within the group stay on topic.
- Foster participation from all participants.
- Foster discussion with open-ended questions.
- Encourage understanding and consensus.
- Discourage disruptive or demeaning behavior.
- Establish ground rules concerning behavioral expectations.
- Establish and maintain a high energy level.
- Use brainstorming and other tools when they add value.
- Keep notes and keep them visible.
- Record the participants' words, not your interpretation.
- Review conclusions and action items.
- Assist the group in smoothly moving through the agenda.

12.4.4 Learning facilitation skills

The following excerpts come from the "FMEA Facilitation Series," which can be found on AccendoReliability.com. Although these excerpts are written for FMEA application, they apply equally well to all types of facilitation.

One of the key factors for successful team leadership is application of specific facilitation skills. The skills needed for excellent facilitation are unique and can be learned. In the words of Ifeanyi Onuoha, "Team leadership is the secret that makes common people achieve uncommon results."

The following are some of the primary facilitation skills that should be mastered to effectively facilitate team meetings to achieve the desired results.

Encouraging participation. Try to gain a balanced involvement and participation from each and every team member, including introverts and extroverts. For information on how to encourage participation, see Carlson (2021a, Facilitation Skill # 1: Encouraging Participation).

Controlling the discussion. Know how to encourage discussion, how to limit discussion, and how to handle someone who dominates the discussion. For information on how to control discussions, see Carlson (2021b, Facilitation Skill #2: Controlling Discussion).

Asking probing questions. Pose direct questions to an individual or to the group to stimulate thinking. These can be used to open up discussion and bring it to a deeper level. For information on how to ask probing questions, see Carlson (2021c, Facilitation Skill #3: Asking Probing Questions).

Asking thought-starter questions. Ask for the elements of FMEA in different ways to help the team think deeply. For information on how to ask thought-starter questions, see Carlson (2021d, Facilitation Skill # 4: Asking Thought-starter Questions).

Active listening. Try to understand thoroughly what another person is saying and why. Practice the advice of Covey (2020): "Seek first to understand, then to be understood." For information on how to actively listen, see Carlson (2022a, Facilitation Skill # 5: Active Listening).

Making decisions. Strive to understand all sides of an issue and find a solution or determine a course of action that is supported by all team members For information on how to make decisions, see Carlson (2022b, Facilitation Skill # 6: Making Decisions).

Managing conflict. Learn the value of disagreements and how to manage them. Remember the words of Mahatma Gandhi, "Honest disagreement is often a good sign of progress." That's why it's called "conflict management." An absence of disagreement can be a cause for concern. For information on managing conflicts, see Carlson (2022c, Facilitation Skill # 7: Managing Conflict).

Brainstorming. Get a flow of ideas on the table before making decisions; this can be most useful when a decision or solution is not easily forthcoming. Albert Einstein and Leopold Infeld (1966, p. 92) said it best:

> The mere formulation of a problem is far more essential than its solution, which may be merely a matter of mathematical or experimental skills. To raise new questions, new possibilities, to regard old problems from a new angle, requires creative imagination and marks real advances in science.

For information on using brainstorming, see Carlson (2022d, Facilitation Skill # 8: Brainstorming).

Fostering creativity. Create an environment that encourages and supports individual and team creativity. For information on fostering team creativity, see the article Carlson (2022e, Creativity and FMEA).

Managing time. Meetings take time and cost money. If time is wasted, meetings will be unsupported and ineffective. Keep meetings focused on what is most important, following the advice of Stephen Covey (2012): "Most of us spend too much time on what is urgent, and not enough time on what is important." For information on how to make meetings more time efficient, see Carlson (2022f, Managing Time) and Section 12.3 in this book.

Getting to consensus. Sometimes, in spite of the best intentions and training, facilitators run into difficulty getting the team to reach consensus. For information and tips on how to achieve consensus in team meetings, see Carlson (2022g, Getting to Consensus).

12.4.5 Individuals, groups, and facilitation

You can and often should get to know the participants before the meeting. When preparing for an important meeting, a facilitator may meet with each invited person prior to the meeting. These one-on-one discussions may involve any preparation or reading the individual should accomplish, along with any of their suggestions, ideas, concerns, etc.

Once the individuals join the group, you will notice another entity has joined the session: the group. The group, as a composite of the individuals, will behave separately from what you may expect from any of the individuals. The group's experiences, responses, and behavior are an element of your facilitation that you need to monitor and accommodate.

By working to create a safe environment where ideas and concepts in service of the objectives flourish, the group will do its best work. Great facilitators recognize that they cannot control outcomes created by the group yet they can establish an environment where the group is best able to accomplish its goals. Facilitators should exert some control to minimize disruptive or inappropriate behavior, guide the meeting to stay on planned focus areas, and work to prevent lengthy, off-topic discussions.

12.4.6 Learning more about facilitation

Here are some excellent articles, proceedings, and books:

The Seven Separators of Facilitation Excellence (http://www.inifac. org/articles/ARSEPAR.pdf, https://www.leadstrat.com/tuesdays-master-facilitation-tip-the-seven-separators-of-facilitation-excellence/) (Wilkinson 2022)

The Secrets of Facilitation by Michael Wilkinson (https://www.mind-tools.com/pages/article/RoleofAFacilitator.htm) (Wilkinson 2012)

Tarmizi, Halbana, Gert-Jan de Vreede, and Ilze Zigurs, Ilze. "Identifying Challenges for Facilitation in Communities of Practice," Proceedings of the 39th Hawaii International Conference on System Sciences (Tarmiz et al. 2006)

Freshley, Craig. "Characteristics of Good Meeting Facilitators." Good Group Decisions. January 15, 2014 (https://www.goodgroupdecisions. com/wp-content/uploads/2013/04/Handout-Characteristics-of-good-meeting-facilitators-2.pdf) (Freshley 2014)

"Basic Facilitation Skills." May 2002. The Human Leadership and Development Division of the American Society for Quality, The Association for Quality and Participation, and The International Association of Facilitators (Burke et al. 2002)

"Community of Practice Design Guide A Step-by-Step Guide for Creating Collaborative Communities of Practice" Copyright 2004, iCohere, Inc. (iCohere 2004)

First Things First by Steven R. Covey et al. (1996)

The above characteristics of well-run meetings and examples of meeting "norms" are from the book *Facilitating with Ease!* written by Ingrid Bens

(2000). This book provides an excellent high-level resource for understanding facilitation methods and skills.

Peter Block's *Flawless Consulting* (2000) is a practical and useful consulting guidebook that focuses on specific consultant skills and behaviors.

Stephen Covey's *7 Habits of Highly Successful People* (2020) provides detailed advice about the habits for successful living and can be easily applied to the field of reliability.

12.4.7 Summary

Every reliability engineer who aspires to build credibility and influence skills should learn how to lead meetings and teams. Becoming a successful facilitator is a path that takes time and practice. Even someone who tends to be shy or introverted can become a proficient facilitator. Fortunately, the skills of effective facilitation can be practiced and learned. This may sometimes require stepping outside one's comfort zone and being willing to make mistakes and listen to feedback. It is well worth the effort.

12.5 Change management

The process to design and deliver a reliable product involves identifying risks. Taking action to understand or mitigate those risks involves much of the day-to-day work of reliability engineering. Taking action to set expectations and improve decisions involves change: change of understanding, change of specifications, change of expectations, and change of designs, processes, and results. It is the changes, big and small, that achieve the desired results for the customer and organization. Obviously, not every suggestion is greeted warmly, not every proposal is funded, and not every recommendation is accepted. Learning to adeptly address these obstacles is crucial to managing change.

12.5.1 Expect resistance to change

When confronted with a detour sign on your normal route to work, you may feel your own resistance to change. You experience a jolt that you need to use an alternate route. Your listing of concerns range from being late for your first meeting to finding an alternate coffee shop.

Adapting to change takes effort. Instead of doing what we are comfortable with, we have to try something else. We already have enough to learn each day and having one more task to master is often not a welcome addition. Think of the work involved to improve the reliability of your system. There is inertia. The team may even recognize the need to improve reliability, yet the looming unknown amount of change may overwhelm those initial steps toward investigating and implementing the proposed changes.

12.5.2 Evidence of negative bias when facing change

A study by Ed O'Brien and Nadav Klein (2017) suggests a negative bias when one is evaluating the data suggesting something has changed. That is, we are much more likely to agree with and accept, as a fact, a set of data showing a negative change, whereas we are less likely to accept as convincing that same magnitude of data for a positive change. Thus, when presenting improvements to product reliability performance, it takes more convincing data to convince others to accept that the improvement is real. A decrease, of the same magnitude, in reliability performance quickly finds acceptance that the changes are significant.

This negative bias exists with actual evidence. Likely, when making recommendations or proposals, the perceived results of the changes likewise face a negative headwind. It is easier to believe failure over success given the same degree of actual change.

12.5.3 Best practices to manage change

Adapting to change requires effort beyond technical assessments, reviews, engineering, evaluations, testing, etc. It takes effort to maintain support across your team to implement and accept the proposed changes. Here are a few best practices to keep in mind as you work to implement change:

Create a clear vision of the benefits of the change. Articulate a concise, crystal clear, understandable summary of the outcome of the proposed change. "Start with the end in mind" (Covey 2020) to provide direction and goals for others to rally around.

Define who does what concerning the proposed change. Who has oversight, makes decisions, implements tasks, enjoys the benefits, etc. Define communication structures, decision authorities, roles and responsibilities, and stakeholders.

Communicate, communicate, and communicate. Communication needs to occur across all the teams involved, including the leadership or oversight team and the stakeholders. Milestones, triggers, and updates all help to keep the project visible, supported, and moving toward the goal.

Find and support the advocates. Some individuals will become champions who actively want to ensure that the project succeeds. Help them with information, insights, and opportunities to share their enthusiasm.

Check assumptions, monitor progress, and review the plan. Setting up a team at the start of a change process is just the start. Help keep your team on the path toward the objective of the change. Likewise, evaluate whether the initial goal remains the right goal.

Support buy-in and support across the organization. Listen and respond honestly, especially to stakeholders, by addressing their concerns and suggestions. Those involved or impacted by the change will have a different view of the situation and impact than you. There is tremendous value in the information they share with you, and by really listening and understanding both the positive and negative comments, you strengthen the chance of a successful result of the change effort.

The following brief story of Carl's illustrates soliciting and getting change in a company:

> Years ago, I was working as a reliability manager at a large original equipment manufacturer and wanted to get the engineering departments more supportive of reliability initiatives. It felt like I was "pushing on a rope." I knew that reliability occurred through the cumulative actions of the entire organization. With my boss's OK, I walked into the Engineering Executive Director's office and asked for his support to engage and inspire the product design departments to achieve higher reliability across the board.
>
> A dozen joint presentations with my Executive Director and me ensued, demonstrating how each department can take steps to achieve reliability in design. We got more support than we anticipated. Sometimes change occurs by stepping out of your normal comfort zone.

Company culture is important. Consider a variant of Newton's first law, the principle of inertia. People will continue to do what is in their personal "comfort zone." Old tools that are not value-added will tend to endure continued use. They have inertia that is difficult to overcome. The introduction of new tools that may add value encounters resistance. As with Newton's third law (i.e., for every action, there is an equal and opposite reaction), it takes strong action from both management and employees to implement positive change and overcome this resistance.

12.5.4 Learning more about change management

Change management is a complex process and well documented in books, courses, seminars, etc. Here are just a few recommended resources, starting with two online pieces on best practices:

Prosci (2022) for 20 years has researched change management and has identified seven factors that consistently are best practices.

10 Best Practices and Advice for the Change Management Process by Jessica Hawley (2017) provides an overview of change management and expands on the 10 best practices.

Abudi (2013) presented a paper titled in part "Change management would be easy if people weren't involved."

12.5.5 Summary

Implementing a plan to achieve a goal or vision will require changes: changes to how we do what we do, the methods we use, or the information considered. Proposed changes nearly always encounter resistance. Hence you should be prepared by thinking through how to manage change. Change management will draw on all the previous skills described in this chapter.

12.6 Summary

Learn to routinely improve each of these soft and related skills as you work in service of your teammates, organization, and customers. Being an effective communicator not only benefits your ability to craft and execute a reliability plan but it also benefits your career as you become seen as an effective leader.

The concept of "lifelong learning" comes to mind as the best approach to mastering your interpersonal and communication skills. We all can improve. Start with where you are by examining what is working or needs improvement. Ask others to assist in your assessment of your skills. Then regularly work to make improvements, get feedback, and continue to hone your skills. It does take time and deliberate practice yet enhances your ability to lead your organization as it transforms its reliability culture.

APPENDIXES

Given that the details in the appendixes continue to evolve, plus the desire to shorten the overall length of the printed work, you may find the appendixes online at Accendo Reliability (accendoreliability.com/go/pre/).

Appendix A: List of Generic Questions to Consider When Conducting a Gap Assessment

This appendix lists potential questions to consider when performing a reliability gap assessment. This list is not meant to be a template and should only be used as thought-starters. The list of questions used in a gap assessment should be suitable for the circumstances and scope of the individual project.

You can find the list of thought-starters for use when preparing to conduct the interviews during a gap assessment at accendoreliability.com/gap-assessment-questions/.

Appendix B: Description of Decision Types

As an aid when selecting methods intended to inform key decisions, we have found that we generally face six different types of questions:

1. Prevention: What can we do now to avoid failures or improve reliability?

2. Comparison: Which design, vendor, or procedure option is better considering reliability?

3. Priority: Where should we focus our resources to best improve reliability?

4. Resources: Who and when should accomplish a specific task?

5. Objective: How do we set or identify the reliability and availability performance objectives, goals, or requirements?

6. Measurement: What is the reliability performance now or expected to be in the future?

Find detailed discussion of each of the decision types at accendoreliability.com/decision-types/.

Appendix C: Reliability Methods Sorted by Method Category

This appendix provides a listing of reliability methods sorted by category with decision type annotated. Keep in mind that this is not an exhaustive list of tools, techniques, or methods but a subset thereof that may be useful as you create a reliability plan. It is a list to provide awareness of common methods that aid in the creation of both a reliability plan and a highly reliable product.

Each listed method includes a very brief description of the method and the typical output. We have found many references for how to execute each of these methods, yet few detail why one would use a method. Therefore, we mention what each method provides to assist in matching the best method for your plan's needs.

Find the list of methods at accendoreliability.com/reliability-methods/.

REFERENCE LIST

For those reading a printed copy, you may find this list online with active links. Find the online reference list at accendoreliability.com/reference-list/.

Abudi, Gina. 2013. "Change Management Would Be Easy If People Weren't Involved: Best Practices for Managing the People-side of Change Management." Project Management Institute, Global Congress 2013, New Orleans, LA.

Anderson, David M. 2020. *Design for Manufacturability: How to Use Concurrent Engineering to Rapidly Develop Low-Cost, High-Quality Products for Lean Production*, 2nd ed. New York: Routledge/Productivity Press.

Arnum, Eric. 2020. "Seventeenth Annual Product Warranty Report." *Warranty Week*, April 16, 2020. https://www.warrantyweek.com/archive/ww20200416.html.

Barr, Chris. 2010. *The Yahoo! Style Guide: The Ultimate Sourcebook for Writing, Editing, and Creating Content for the Digital World.* New York: St. Martin's Griffin.

Bens, Ingrid. 2000. *Facilitating with Ease!* San Francisco: Jossey-Bass.

Block, Peter. *Flawless Consulting: A Guide to Getting Your Expertise Used.* San Francisco: Jossey-Bass, 2011.

Bollinger, Terry B., and Clement McGowan. 1991. "A critical look at software capability evaluations." *IEEE Software* 8(4): 25–41.

Bradin, John S. 1988. "Organizing and Managing the Reliability Function." In *Handbook of Reliability Engineering and Management*, 1st ed., edited by W. Grant Ireson and Clyde F. Coombs Jr., 2.3–2.36. New York: McGraw-Hill. p. 2.9.

Brombacher, Aarnout C. 1999. "Maturity index on reliability: Covering non-technical aspects of IEC61508 reliability certification." *Reliability Engineering & System Safety* 66(2): 109–120.

Brooks, Alison Wood, and Leslie K. John. 2018. "The surprising power of questions." *Harvard Business Review* 96(3): 60–67. https://hbr.org/2018/05/the-surprising-power-of-questions.

Bryan, Mark, Julia Cameron, and Catherine Allen. 1998. *The Artist's Way at Work: Riding the Dragon, Twelve Weeks to Creative Freedom*. New York: William Morrow.

Burke, Dennis W., Melanie Donahoe, Rudolph Hirzel, Linda Mather, Gail Morgenstern, Ned Ruete, Ed Smith, Deborah Starzynski, and Jo Ann Stoddard, contributors. 2002. Basic Facilitation Skills. Human Leadership and Development Division of the American Society for Quality, Association for Quality and Participation, and International Association of Facilitators. May 2002. https://my.asq.org/communities/files/120/1188.

Carlson, Carl S. 2021a. "Facilitation Skill #1: Encouraging Participation," Inside FMEA (blog), Accendo Reliability. September 1, 2021. https://accendoreliability.com/facilitation-skill-1-encouraging-participation/.

Carlson, Carl S. 2021b. "Facilitation Skill #2: Controlling Discussion," Inside FMEA (blog), Accendo Reliability. October 1, 2021. https://accendoreliability.com/facilitation-skill-2-controlling-discussion/.

Carlson, Carl S. 2021c. "Facilitation Skill #3: Asking Probing Questions," Inside FMEA (blog), Accendo Reliability. November 1, 2021. https://accendoreliability.com/facilitation-skill-3-asking-probing-questions/.

Carlson, Carl S. 2021d. "Facilitation Skill #4: Asking Thought-starter Questions," Inside FMEA (blog), Accendo Reliability. December 1, 2021 https://accendoreliability.com/asking-thought-starter-questions/.

Carlson, Carl S. 2022a. "Facilitation Skill #5: Active Listening," Inside FMEA (blog), Accendo Reliability. January 1, 2022. https://accendoreliability.com/facilitation-skill-active-listening/.

Carlson, Carl S. 2022b. "Facilitation Skill #6: Making Decisions," Inside FMEA (blog), Accendo Reliability. February 1, 2022. https://accendoreliability.com/facilitation-skill-6-making-decisions/.

Carlson, Carl S. 2022c. "Facilitation Skill #7: Managing Conflict," Inside FMEA (blog), Accendo Reliability. January 1, 2022. https://accendoreliability.com/facilitation-skill-7-managing-conflict/.

Carlson, Carl S. 2022d. "Facilitation Skill #8: Brainstorming," Inside FMEA (blog), Accendo Reliability. January 1, 2022. https://accendoreliability.com/facilitation-skill-8-brainstorming/.

Carlson, Carl S. 2022e. "Creativity and FMEA," Inside FMEA (blog), Accendo Reliability. May 1, 2022. https://accendoreliability.com/creativity-and-fmea/.

Carlson, Carl S. 2022f. "Managing Time," Inside FMEA (blog), Accendo Reliability. July 1, 2022. https://accendoreliability.com/managing-time/.

Carlson, Carl S. 2022g. "Unique Challenges When Facilitating FMEAs," Inside FMEA (blog), Accendo Reliability. June 1, 2022. https://accendoreliability.com/common-fmea-facilitation-problems/.

Collins, James C., and Jerry I. Porras. 1996. "Building your company's vision." *Harvard Business Review* 74(5): 65.

Communication in the Real World: An Introduction to Communication Studies. 2016. Section: "5.3 Improving Listening Competence." Open

Textbook Library. Minneapolis: University of Minnesota Libraries. 29 September 2016. doi:10.24926/8668.0401. ISBN 9781946135070. OCLC 953180972. https://open.lib.umn.edu/communication/chapter/5-3-improving-listening-competence/.

Conaway, Cameron. 2022. "The right way to process feedback" *Harvard Business Review* June 14, 2022. Ascend. Harvard Business Review. https://hbr.org/2022/06/the-right-way-to-process-feedback. Updated on 14 June 2022.

Covey, Stephen R. 2012. *The Wisdom and Teachings of Stephen R. Covey*. New York: Free Press.

Covey, Stephen R. 2020. *The 7 Habits of Highly Effective People*, 30th Anniversary ed. New York: Simon & Schuster. May 19, 2020.

Covey, Stephen R., A. Roger Merrill, and Rebecca R. Merrill. 1995. *First Things First*, reprint edition. New York: Free Press.

Crosby, Philip B. 1980. *Quality Is Free: The Art of Making Quality Certain*. New York: Mentor.

Deming, W. Edwards. 2000. *Out of the Crisis*, reprint ed.. Cambridge, MA: The MIT Press.

Deutschendorf Harvey. 2014. "5 Ways to Have Great Conversations." *Fast Company*, https://www.fastcompany.com/3027801/5-ways-to-have-great-conversations. Posted 24 March 2014.

Doran, George T. 1981. "There's a S.M.A.R.T. way to write management's goals and objectives." *Management Review* 70(11): 35–36.

Duarte, Nancy. 2008. *Slide:ology: The Art and Science of Creating Great Presentations*. Sebastapol, CA: O'Reilly Media.

Duarte, Nancy. 2013. *Resonate: Present Visual Stories That Transform Audiences*. Hoboken, NJ: Wiley.

Duarte, Nancy, and Patti Sanchez. 2016. *Illuminate: Ignite Change through Speeches, Stories, Ceremonies, and Symbols.* New York: Penguin.

edX team. 2022. "Tips for Successful Online Learning." edX Blog. edX. https://blog.edx.org/tips-for-successful-online-learning. Accessed 27 July 2022.

Einstein, Albert, and Leopold Infeld. 1966. *The Evolution of Physics: From Early Concepts to Relativity and Quanta*, new ed. New York: Simon & Schuster.

Expert Academy. 2021. "Asking Questions—Types & Examples." October 28, 2021. Video, 2:47. https://youtu.be/bkelnH9SWNw.

Freshly, Craig. 2014. "Characteristics of Good Meeting Facilitators." PDF download. Good Group Decisions. https://www.goodgroupdecisions.com/wp-content/uploads/2013/04/Handout-Characteristics-of-good-meeting-facilitators-2.pdf. January 15, 2014.

Froke, et. al. 2020. *The Associated Press Stylebook*, 55th ed. New York: Basic Books.

Fruehling, Rosemary T., and Neild B. Oldham. 1988. *Write to the Point!: Letters, Memos, and Reports That Get Results.* New York: McGraw-Hill.

Gallo, Carmine. 2016. *The Storyteller's Secret: From TED Speakers to Business Legends, Why Some Ideas Catch on and Others Don't.* New York: St. Martin's Press.

Gnanapragasam, Alex, Christine Cole, Jagdeep Singh, and Tim Cooper. 2018. "Consumer perspectives on longevity and reliability: A national study of purchasing factors across eighteen product categories," *Procedia CIRP* 69(201): 910–915. https://doi.org/10.1016/j.procir.2017.11.151.

Grinder, Michael. 2022. Group Wizardry. Michael Grinder & Associates. Accessed 28 June 2022. https://michaelgrinder.com/upcoming-events/group-wizardry/.

Harvard Business Review. 2015. "The Art of Asking Questions." Video, 1:14. https://hbr.org/video/4457382113001/the-art-of-asking-questions.

Hawley, Jessica. 2017. "10 Best Practices and Advice for the Change Management Process." Quickbase (blog). http://www.quickbase.com/blog/10-best-practices-in-change-management. Posted 9 March 2017.

iCohere. 2004. "Community of Practice Design Guide: A Step-by-Step Guide for Creating Collaborative Communities of Practice." iCohere, National Learning Infrastructure Initiative at EDUCAUSE, and American Association for Higher Education. https://www.abcee.org/sites/default/files/communities_of_practice_design_guide_0.pdf.

IEEE Standards Board. 2009. "IEEE Standard for Organizational Reliability Capability," IEEE Std 1624-2008, New York: IEEE. 5 February 2009.

Jeary, Tony. 2004. Life Is a Series of Presentations: 8 Ways to Punch up Your People Skills at Work, at Home, Anytime, Anywhere. New York: Simon & Schuster.

Koshy, Kiron, Alison Liu, Katharine Whitehurst, Buket Gundogan, and Yasser Al Omran. 2017. "How to hold an effective meeting." International Journal of Surgery: Oncology 2(5): e22. https://www.ncbi.nlm.nih.gov/pmc/articles/PMC5916468/.

Lane, Andrew Stuart, and Chris Roberts. 2022. "Contextualize reflective competence: A new learning model promoting reflective practice for clinical training" BMX Medical Education 22:71, published online 30 January 2022. doi:10.1186/s12909-022-03112-4. https://www.ncbi.nlm.nih.gov/pmc/articles/PMC8801113/.

Levitt, D.H. 2001. "Active listening and counselor self-efficacy: Emphasis on one microskill in the beginning counselor training." *The Clinical Supervisor* 20(2): 101–115. doi:10.1300/J001v20n02_09. S2CID 145368181.

Markel, Mike, and Stuart A. Selber. Markel. 2021. *Technical Communication*. Boston: Macmillan.

Marshall, Lisa J., and Lucy D. Freedman. 1995. *Smart Work: The Syntax Guide for Mutual Understanding in the Workplace*. Dubuque, IA: Kendall/Hunt.

McKean, Erin. 2005. *The New Oxford American Dictionary*, 2nd ed. Oxford: Oxford University Press.

Mind Tools Content Team. 2022. MindTools, Communication Skills. Accessed 22 July 2022. https://www.mindtools.com/CommSkll/ActiveListening.htm.

Morgan, Kieran. 2015. *Technical Writing Process: The Simple, Five-step Guide That Anyone Can Use to Create Technical Documents Such as User Guides, Manuals, and Procedures*. Better on Paper.

Moss, Richard W. II. 1996. "Design for Reliability." In *Handbook of Reliability Engineering and Management*, 2nd ed., edited by W. Grant Ireson, Clyde F Coombs, and Richard Y Moss. 5.1–5.16. New York: McGraw Hill.

O'Brien, E., & Klein, N. 2017. "The tipping point of perceived change: Asymmetric thresholds in diagnosing improvement versus decline." *Journal of Personality and Social Psychology* 112(2): 161–185. https://doi.org/10.1037/pspa0000070.

O'Connor, Patrick, and Andre Kleyner. 2012. *Practical Reliability Engineering*, 5th ed. Chichester, UK: Wiley.

Oakley, Barbara, and Terrence Sejnowski. 2022a. "Learning How to Learn: Powerful Mental Tools to Help you Master Tough Subjects." Coursera. https://www.coursera.org/learn/learning-how-to-learn. Accessed 27 July 2022.

Oakley, Barbara, Dr. Terrence Sejnowski. 2022b. "Mindshift: Break through Obstacles to Learning and Discover Your Hidden Potential." Coursera https://www.coursera.org/learn/mindshift. Accessed 27 July 2022.

Oakley, Barbara, and Terrence Sejnowski. 2022c. "Uncommon Sense Teaching: Teaching Online." Coursera. https://www.coursera.org/learn/teaching-online. Accessed 26 September 2022.

Perry, Elizabeth. 2022. "Good Question. Impress Everyone with Our Guide to Asking Better Ones. BetterUp.com. Last updated on February 21, 2022. https://www.betterup.com/blog/how-to-ask-good-questions.

Petroski, Henry. 1994. *Design Paradigms: Case Histories of Error and Judgment in Engineering.* Cambridge: Cambridge University Press.

Prosci. 2022. "Best Practices in Change Management." Prosci (blog). https://www.prosci.com/resources/articles/change-management-best-practices/. Accessed 12 June 2022.

Rathore, Shyamli. 2022. "How to Lead Better Virtual Meetings." Ascend newsletter. *Harvard Business Review.* https://hbr.org/2022/07/how-to-lead-better-virtual-meetings. Accessed 5 July 2022.

Repenning, Nelson P., and John D. Sterman. 2001. "Nobody ever gets credit for fixing problems that never happened: Creating and sustaining process improvement." *California Management Review* 43(4): 64–88. https://doi.org/10.2307/41166101.

Repenning, Nelson P., Paulo Goncalves, and Laura J. Black. 2001. "Past the tipping point: The persistence of firefighting in product development." *California Management Review* 43(4): 44–63.

Rogelberg, Steven G. 2019. "Why your meetings stink and what to do about it." *Harvard Business Review* 97(1): 140–143.

Rosling, Hans. 2022. "The best Hans Rosling talks you've ever seen." TED Ideas Worth Spreading. https://www.ted.com/playlists/474/the_best_hans_rosling_talks_yo. Accessed 3 August 2022.

Schenkelberg, Fred. 2014. *Finding Value: How to Determine the Value of Reliability Engineering Activities.* Los Gatos, CA: FMS Reliability Publishing.

Schenkelberg, Fred. 2016. "Reliability Management." In *Quality and Reliability Management and Its Applications*, edited by Hoang Pham, 309–351. London: Springer-Verlag. doi:10.1007/978-1-4471-6778-5_11.

Schenkelberg, Fred. 2017 "Creating meaningful derating graphics." In *2017 Annual Reliability and Maintainability Symposium (RAMS)*, pp. 1–6. IEEE.

Schenkelberg, Fred. 2022a. "The Basics of Derating Electronic Components." Musings on Reliability and Maintenance Topics (blog). Accendo Reliability. https://accendoreliability.com/basics-derating-electronic-components/. Accessed 7 April 2022.

Schenkelberg, Fred. 2022b. "The 7 Essential Elements of a Complete Request." Musings on Reliability and Maintenance Topics (blog). Accendo Reliability. https://accendoreliability.com/7-essential-elements-complete-request/. Accessed 10 April 2022.

Schilling, Dianne. 2012. "10 steps to effective listening." *Forbes.* https://www.forbes.com/sites/womensmedia/2012/11/09/10-steps-to-effective-listening. Accessed 12 June 2022.

Schweitzer, Maurice. 2022. "Improving Communication Skills." Wharton University of Pennsylvania. Coursera. Accessed 23 June 2022. https://www.coursera.org/learn/wharton-communication-skills.

Senge, Peter M. 1990. *The Fifth Discipline: The Art & Practice of The Learning Organization*. New York: Doubleday Currency.

Shukla, Aditya. 2021. "What's the Best Way to Learn Anything? 16 Tips from Science. Cognition Today Inside Your Mind (blog). https://cognitiontoday.com/best-ways-to-learn-anything-efficiently/.

Simmons, Annette. 2015. *Whoever Tells the Best Story Wins: How to Use Your Own Stories to Communicate with Power and Impact*. New York: Amacom.

Simon, Carmen. 2016. *Impossible to Ignore: Creating Memorable Content to Influence Decisions*. New York: McGraw Hill Professional.

Simon, Carmen. 2022a. "The Neuroscience of Digital Content." Corporate Visions. Video. Accessed 15 July 15 2022. https://corporatevisions.com/resources/memorable-marketing/webinar-replay-neuroscience-digital-content/.

Simon, Carmen. 2022b. "Deliver Memorable Virtual Presentations." Marketo part of Adobe Experience Cloud. Video. Accessed 15 July 15 2022. https://www.marketo.com/webinars/deliver-memorable-virtual-presentations/.

Skillin, Marjorie, and Robert Gay. 1974. *Words into Type*, 3rd ed. Hoboken, NJ: Prentice Hall.

Slayler, Sharon. 2011. *What Your Body Says (and How to Master the Message): Inspire, Influence, Build Trust, and Create Lasting Business Relationships*. Hoboken, NJ: Wiley.

Strunk Jr., William, and Elwyn Brooks White. 1979. *The Elements of Style*, 3rd ed. New York: MacMillan.

Tarmizi, Halbana, Gert-Jan de Vreede, and Ilze Zigurs. 2006. "Identifying Challenges for Facilitation in Communities of Practice." In *Proceedings of the 39th Annual Hawaii International Conference on System Sciences* (HICSS'06), vol. 1, p. 26a. New York: IEEE.

Tufte, Edward R. 1990. *Envisioning Information*. Cheshire, CT: Graphics Press.

Tufte, Edward R. 1997a. *Visual and Statistical Thinking: Display of Evidence for Decision Making*. Cheshire, CT: Graphics Press.

Tufte, Edward R. 1997b. *Visual Explanations: Images and Quantities, Evidence and Narrative*. Cheshire, CT: Graphics Press.

Tufte, Edward R. 2001. *The Visual Display of Quantitative Information*, 2nd ed. Cheshire, CT: Graphics Press.

Tufte, Edward R. 2006. *Beautiful Evidence*. Cheshire, CT: Graphics Press.

Tufte, Edward R. 2020. *Seeing with Fresh Eyes: Meaning, Space, Data, Truth*. Cheshire, CT: Graphics Press.

University of Chicago Press. 2010. *Chicago Manual of Style*, 16th ed. Chicago: The University of Chicago Press.

US Military Standard [MIL-STD]-785B. 1980. *Reliability Program for Systems and Equipment Development and Production*. Military Standard. Washington DC: Department of Defense. (Note: Status Canceled as of 30 July 1998.)

Van Laan, Krista. 2012. *The Insider's Guide to Technical Writing*. Denver: XML Press.

Vizabi. 2022. "Interactive Charts and Visualization Tools Animated through Time." Vizabi v0/25.1-3. https://vizabi.org. Accessed 12 August 2022.

Wikipedia. 2022. "Active Listening." Accessed 8 July 8 2022. https://en.wikipedia.org/wiki/Active_listening.

Wilkinson, Michael. 2012. *The Secrets of Facilitation: The SMART Guide to Getting Results with Groups*. New York: Wiley.

Wilkinson, Michael, 2022. "The Seven Separators of Great Facilitation." Leadership Strategies (blog). Accessed August 19, 2022. https://www.leadstrat.com/tuesdays-master-facilitation-tip-the-seven-separators-of-facilitation-excellence/.

Wilson, Pamela. 2016. *Master Content Marketing: A Simple Strategy to Cure the Blank Page Blues and Attract a Profitable Audience*. Nashville, TN: BIG Brand Books.

Wilson, Pamela. 2018. *Master Content Strategy: How to Maximize Your Reach and Boost Your Bottom Line Every Time You Hit Publish*. Nashville, TN: BIG Brand Books.

Yoshimura, Tatsuhiko. "Why you couldn't succeed in lean development. Toyota style GD3 & APAT are the keys for lean development." Presented at KLMANAGEMENT, France, date unknown. Accessed on 1 August 2022. http://www.klmanagement.fr/images/PDF/Articles/3%20GD3%20and%20APAT%20are%20the%20Keys%20for%20LD-Yoshimura%20JMAC%20Tokyo-KLMANAGEMENT.pdf.

ABOUT THE AUTHORS

Carl S. Carlson

Carl is a consultant and instructor in the areas of Failure Mode and Effects Analysis and other reliability and quality disciplines, supporting clients from a wide cross section of industries. He has 40 years of experience in reliability testing, engineering, and management positions, including as senior manager for advanced reliability at General Motors. He has a Bachelor's degree in mechanical engineering from the University of Michigan, is a Senior Member of the American Society of Quality, and a Certified Reliability Engineer. His book, Effective FMEAs, was published by John Wiley & Sons in 2012, and he regularly writes and podcasts on AccendoReliability.com.

Fred Schenkelberg

Fred is an international authority on reliability engineering and founder of Accendo Reliability. He is the reliability expert at FMS Reliability, a reliability engineering and management consulting firm he founded in 2004. Fred left Hewlett Packard's Reliability Team where he helped create a culture of reliability across the corporation to assist other organizations. His passion is working with teams to improve product reliability, customer satisfaction, and efficiencies in product development and to reduce product risk and warranty costs. Fred has a Bachelor of Science in physics from the United States Military Academy and a Master of Science in statistics from Stanford University.

www.ingramcontent.com/pod-product-compliance
Lightning Source LLC
Chambersburg PA
CBHW071551210326
41597CB00019B/3193